Environmental Threats

Environmental Threats: Perception, Analysis and Management

Edited by
Jennifer Brown

Published in association with

ECONOMIC AND SOCIAL
RESEARCH COUNCIL
ESRC

Belhaven Press
A division of Pinter Publishers
London and New York

First published in Great Britain in 1989 by
Belhaven Press (a division of Pinter Publishers),
25 Floral Street, London WC2E 9DS

British Library Cataloguing in Publication Data

A CIP catalogue record for this book is available from the British Library

ISBN 1 85293 015 2

Library of Congress Cataloging-in-Publication Data

Environmental threats.

Based on papers prepared for a July 1986 seminar organized by the Dept. of the Environment and the Economic and Social Research Council.

"Published in association with Economic and Social Research Council."

Includes index.

1. Environmental impact analysis — Congresses. 2. Risk management — Congresses. I. Brown, Jennifer, 1948- . II. Great Britain. Dept. of the Environment. III. Economic and Social Research Council (Great Britain)
TD194.6.E67 1989 333.7'15 88-34235

ISBN 1-85293-015-2

Filmset by Mayhew Typesetting, Bristol, England
Printed by Biddles Ltd, Guildford and King's Lynn

Contents

List of Illustrations

Figures

Photographs *between pages 66-7*

Tables

Preface

The study of risk perception is particularly suited to the social sciences. Risk is normally defined as the combination of an event (or a pathway) with the potential to cause injury or death, and the likelihood of occurrence. Three types of uncertainty are evident: that surrounding the possible outcome; that involved in the estimation of probability and perhaps the most intriguing, that applied to people's judgements and reactions to these two bands of indeterminacy. No two individuals will respond in the same way. Some will see risk as a challenge, others a symbol of a disliked and distrusted technology or managing agency. Some will regard risk as an act of God, even when it is manifestly human-caused, others will look for a culprit even when the reason for the hazard is unknown.

Social scientists have become fascinated with public attitudes towards risk because it is so evident that technical valuations alone provide an insufficient guide to regulation. No matter how much the specialists sneer at an 'irrational' and 'ignorant' public, lay judgements about possible dangers are equally as important as scientific or technical analysis. Indeed in some cases, such as the regulation of persistent toxic chemicals and radioactive releases, popular interpretations dominate political decisions, no matter how much the technical advisers protest that safety standards are unrealistically expensive and safety margins too high.

The Economic and Social Research Council (ESRC) sponsored an Environmental Issues Initiative between 1985 and 1987. This consisted of a number of major research projects and the intellectual output of four working groups. These groups consisted of academics, practitioners, regulatory officials and other interested parties who enjoyed discussing a variety of themes around four topics. One of these topics was risk perception. Such is the political interest in this area that the Department of the Environment collaborated with the ESRC over two one-day seminars in July 1986 and May 1987. This book is the product of the papers prepared for the 1986 meeting by the Risk Working Group.

The result is a stimulating blend of creative thinking and practical case studies. Many of the ideas expressed in the pages that follow would not have been generated had the Working Group not been so committed and enthusiastic over its two years of existence. This is a tribute to the abilities and perseverance of the authors and the hard-working editor. Good analysis always creates a need for further, and even more creative research. The vexed question of risk communication needs detailed treatment through imaginative experiments. The scope for revealing common ground underlying the judgements of the various groups of 'risk' professionals is also crying out for exploration through appropriate case studies. The improvement of processes and institutions designed to assess levels of safety and to command public trust and support in the outcome is a matter of urgency notably with regard to the disposal of radioactive material and toxic chemicals.

The Risk Working Group has made a bold and imaginative start to the work that must now be done. They are to be congratulated for their efforts and for their leadership.

Timothy O'Riordan
University of East Anglia

Foreword

As a consequence of the United Kingdom's Economic and Social Research Council's environmental issues initiative, the risk group was one of four convened to examine environmental topics in need of a multidisciplinary approach. The core group was comprised of Silvio Funowicz, Alan Irwin, Sally Macgill, Jerry Ravetz, Michael Thompson and Brian Wynne. In addition the group benefited from contributions of corresponding members: Jennifer King, Erik Millstone, Derek Morgan and Colin Mackay.

Early in the life of the group three broad themes were identified for special attention:

(1) the influence and role of the managing institutions within which policy and decisions about potential or actual environmental hazards are made;
(2) an evaluation of the effectiveness of methods or tools used in risk assessment and their contribution to policy and decision-making;
(3) the issues of risk perception as seen by the public and by experts.

In addition, the focus of interest was largely concerned with technological risks and the threats these posed.

The risk group wished to emphasise practical issues and to provide an accessible account of social science analyses of the reasons underpinning the different views held by public and experts over threats to the environment. It was concluded that much could be explained by examining the psychological, social and political realities underlying people's definitions of risks and the solutions they preferred for their control and management. Analysis initially involves the discovery and elaboration of the social processes inherent in the selection of risks giving rise to concern followed by the preferred methods by which to avoid any harmful consequences.

The following chapters are offered as a contribution towards understanding the different views, often impassionately held, between those who create, regulate and bear the potential consequences of risks posed by the technological environment.

Jennifer Brown
Convenor, Risk Working Group 1985–7
July 1988

Acknowledgements

The chapter by Peter James and Michael Thompson was originally prepared for a conference held at the University of East Anglia and appears with permission of Professor Lewis Roberts. Jerry Ravetz prepared the material for his contribution to the chapter on biotechnology while a member of the Working Party on the Safety of Genetically Engineered Organisms convened by the Council for Science and Society.

Photographs were supplied by Joyce Henderson, Keith Howard and Ian Walters. Copyright permission to reproduce the other illustrations was given by Nelson-Hall publishers, the Electricity Council and British Nuclear Fuels Ltd. The children's drawing were collected during a study by Jennifer Brown and Joyce Henderson. The help of the participating schools and the children whose drawings are reproduced in the book is gratefully acknowledged.

Members of the Risk Group showed very considerable commitment and industry in developing the papers presented in this book and I gratefully acknowledge their support, patience and tolerance during its production.

Professor Tim O'Riordan, Chair of the environmental issues initiative, Cathy Palmer from the ESRC and Joyce Tait the co-ordinator of the working groups were always considerate and supportive in providing help and advice for the smooth functioning of the Risk Group.

I would also like to express my gratitude to Di Wells, Mary Banbury and Pauline Elliott from the University of Surrey who provided the administration and secretarial back-up for the Group with grace, good humour and skill.

List of contributors

Jennifer Brown, Department of Psychology, University of Surrey.

Silvio Funtowicz, Research Methods Consultancy, Sheffield.

Alan Irwin, Department of Science and Technology Policy, University of Manchester.

Peter James, Department of Business and Management, University of Stirling.

Sally Macgill, School of Geography, University of Leeds.

Jerry Ravetz, Department of Philosophy, University of Leeds.

Michael Thompson, Independent researcher.

Brian Wynne, School of Independent Studies, University of Lancaster.

Chapter 1
Introduction: Approaches, Tools and Perceptions

Jennifer Brown

Threats are intimidatory announcements, actions, or impending evils that are likely to have undesirable consequences. [Oxford English Dictionary]

Threat is a broader term than risk, which more usually is confined to the probability or likelihood of an event occurring and causing harm. Environmental threats are taken to be the potential or actual disruption of the expected functioning of the natural or technological worlds. The focus of this edited collection of papers is those actual or potential threats imposed by science and technology and the efforts to minimise their undesirable or untoward consequences. The aim is to present — in accessible language — social science analyses of the public's, the technical expert's and the policymaker's perspectives concerning their identification of and response to such threats. It is hoped thereby to contribute towards understanding the reasons for public uncertainty and policymakers' puzzlement over the apparent disparity between lay and expert judgements about risk calculations and the attendant distrust and acrimony in attempts to introduce technological innovations.

Background

The activities of environmental pressure groups and the resistance of local communities to hosting potentially hazardous industrial plant are witness to the growing public concern about the impact of technologies on their own health and well-being as well as possible harm to the physical environment. It is perhaps all too easy to caricature the public as having little technical knowledge and demonstrating that their emotional adverse reactions to technical processes are an irrational fear of the unknown, while the expert, armed with knowledge and certainty, coolly and unemotionally appraises the technology in question as having an acceptable level of risk. Issues embedded in this over-simplified proposition are explored in different ways by the present authors. Funtowicz and Ravetz in their chapter and Thompson's contribution examine the issue of (un)certainty; Macgill looks in detail at the logic of the public's perception of radioactive waste discharges, while Wynne and Irwin question the assumptions of scientific knowledge or method as applied to risk assessment. As a prelude to those chapters, the remainder of this introduction sets out the social context of environmental threats and the approaches utilised to analyse specific issues.

Social context of risk

The physical world we live in is and always has been precarious. The elements, notably fire and water, can provide sustaining warmth and power but they can also cause injury and great devastation. Science and technology have exploited the environment largely for the benefit of human beings. The focus of this book is a social science analysis of the threats posed not by the physical world *per se* but by humanly constructed technologies that harness its resources. Kasperson and Pijawka (1985, 7) note that 'technology has emerged as the major source of hazard to modern society'. They make a further important point that hazards deriving from technology set quite different management problems to those of natural hazards. The former are often unfamiliar and lack precedents in methods of control, accidents frequently result in chronic and long-term consequences and there may be many exposed individuals.

The study of risk is a relatively new and burgeoning area of research for social scientists (Canter *et al.*, 1985). Disciplines represented in the present book include sociology, psychology, cultural anthropology, geography, philosophy and policy analysis. These disciplines provide accounts of risks which both challenge assumptions of a purely technical analysis and place risks within a social context that allows interpretation of the considerably raised public profile of actual or potential threats resulting from technology. A rationale for such analyses is provided by Sjoborg (1987, 1).

> Decision makers are faced with the intense concerns of various individuals and groups regarding what they see as potential dangers in the environment, dangers often associated with the development of advanced and complex new technology. The mass media are often quite eager to spread such information about risks and to carry on with the debate. Demands for participation in the decision making process are voiced and politicians and administrators find their sphere of action constrained, caught as they are in the midst of intense social conflict and controversy over risks.

The role of the decision-making institutions, the part played by the media, the different perspectives of individuals and groups are all issues explored by the present authors. They do so with a view to making the underlying reasons for much of the conflict over potential hazardous technology clear and accessible to as wide an audience as possible. The aim is to facilitate understanding and provide a route out of the ever-growing bitterness of clashes between affected publics and the managing institutions.

Mastery of the environment through discovery and application has meant that science and technology have harnessed the benefits and to a large extent offset the harm that the elements can wreck. Better buildings and improvements in public health and sanitation on the one hand and development of the energy, chemical and electronic industries on the other have led to longer, more comfortable lives with people, in the industrialised world at least, enjoying the material benefits of consumer products. Yet paradoxically, while enjoying its fruits public concern about the environment in general and the threat of harm from science and technology in particular is increasing. Douglas and Wildavsky (1982, 51) express this idea as follows: 'in the amazingly short space of 15–20 years confidence about the physical world has turned to doubt. Once the source of safety, science and technology have become the source of risk'. This paradox lies at the heart of perception of the environment as threatening. Kates (1981, 81) presents two arguments. The first is that while people are less likely to die of infectious disease they are more likely to suffer consequences of stress, malignancy and car accidents — i.e. threats that are humanly induced. He continues (83) 'even if the everyday is more secure, the exceptional is less so. Medical advances, improvements in safety, engineering devices to protect against natural hazard, while they prove effective in eliminating much of the day-to-day environmental threats of life, nonetheless serve to increase the magnitude and frequency of the catastrophic'. An alternative view is that the environment is indeed less threatening, it only *appears* more hazardous. This is due to changes in social expectations and perceptions. Whether the environment in objective terms is more threatening because of the scale of technological advance or is only perceived as such, matters little to the thrust of this book. There is a demonstrable climate of opinion that does believe that there are threats to life and property from the technological environment.

The present contributors start with the premise that quantative assessment of risk in purely physical terms is not only impossible, but the exclusion of other qualitative factors by the technocratic community has led to an increasing disillusionment with the 'expert'. The central thesis running through the following chapters is that environmental threats and attendant risks have social meaning and these have to be taken into account if a dialogue is to be enacted between the risk creators and risk bearers. Improving understanding and thus the possibility of effective communication between engineer and social scientist is very much the focus of Thompson's chapter. He argues that this is 'an urgent and far from trivial task' if the tensions between those imposing and bearing technological risks stand any chance of being reduced. In Chapter 3 Wynne cautions that public debate of these issues will 'founder in confusion and bad faith' unless efforts are made to negotiate a more appropriate relationship between experts and public.

None of the present authors would deny that safety and integrity of the physical systems are paramount. That is taken as axiomatic. However, scientific and engineering methodologies while necessary are insufficient to provide an analysis of the impact of technologies on communities. Without the additional layer of understanding that social science analysis can contribute, the gap between expert and public will widen, the financial costs of dispute will escalate and the social costs of polarisation of society will deepen. Irwin and Wynne provide accounts of the manner in which technical and policymaking bodies undertake risk assessments and the assumptions underpinning them. Irwin examines the disagreement between experts over the interpretation of road accident figures following the introduction of seat belt legislation. Wynne, by reference to a fatal explosion of methane gas in an underground valve house at Abbeystead, shows how the affected public operated with very different assumptions to those of investigators from the inquiry team. In part this explains the disappointment and frustration experienced by the victims' families with the officers of the Health and Safety Executive inquiring into the accident's causes.

How do we know that the environment, and more particularly the technological environment, is seen as threatening? There are a number of indicators that contribute to a public climate against which individual judgements are made. These indicators include:

(1) the catalogue of large-scale technological disasters;
(2) the coverage by the mass communication media;
(3) erosion of trust in experts;
(4) the politicisation of environmental issues.

Increase in technological disasters

There have always been accidents following the application of science and developments in technology and have been well charted by Lagadec (1982). In particular, development of transport systems and mining in the early centuries of human history resulted in many thousands of deaths. In later centuries as engineering techniques improved there also evolved a less cavalier attitude towards the work force and the view that disasters were not purely Acts of God but often due to human incompetence or negligence. The nature of twentieth-century technological disasters increased the scale of damage and involved bystanders or the local public.

In the United Kingdom during the 1970s a number of accidents added to the ever-growing catalogue of contemporary disasters. The Summerland Leisure Complex on the Isle of Man caught fire and 53 holidaymakers were killed. The ignition of cyclohexane at Nypro's Flixborough factory profoundly shook industry and shocked the public. Flixborough, and later accidents: Seveso, Three Mile Island, Bhopal and Chernobyl, have become synonymous with large-scale industrial disaster that illustrate not only the frailty of these industries but the enormous costs when things do go wrong. There have also been a number of spectacular pollution disasters such as Love Canal in the United States, the massive oil spillages following the *Torrey Canyon* and *Amoco Cadiz* tanker collisions. Most recently the Sandoz plant near Basle caught fire and severely polluted the Rhine. The differences between the earlier history of technological disasters and these present examples are not just a matter of scale and the involvement of the public as victims but the attention drawn to them by the mass communication media.

Media attention

It is a contentious issue whether the media reflect or initiate public concern, but it is through television and newsprint that the drama of the Zeebrugge ferry disaster or King's Cross underground fire is instantly brought to a mass audience. This has been called the dependency effect (Short, 1984): in other words the reliance on the mass media, most significantly television, for information. Lee (1986b) in reviewing the impact of the media on attitudes towards chemical hazards concludes that there is a dearth of evidence about attitude change following media coverage. It is more likely the media consolidates rather than changes views. However, Mazur (1984) finds a striking correlation between the increasing amount of coverage of the nuclear power controversy in the United States and an increase in opposition to nuclear plant as measured by opinion poll surveys. This analysis proposes that the media have an important agenda-setting role in deciding what issues to cover. The controversies surrounding technology, accidents at industrial plant or the activities of active protestors are newsworthy stories. The nation becomes a 'spectator' to the unfolding drama, or as O'Riordan (1983) suggests scientific or technological projects with their institutional complications are the focus of attention for 'political theatre'.

The sense in which scientific disputes become the focus for media attention is described by Macgill in her chapter on the controversy surrounding radioactive discharges from British Nuclear Fuels' reprocessing plant in Sellafield, Cumbria. In particular the report by Sir Douglas Black on the hypothesised link between the Sellafield discharges and local incidence of childhood leukaemia and other cancers is discussed in relation to the treatment of the report's findings by the media and their impact on the local population. Sir Douglas Black's status as an impartial expert was a significant feature in the interpretation and credibility of the results of his investigation. Experts increasingly and uncomfortably find themselves the subject of media attention. Reported incompetence, negligence, error or misdeeds have contributed to the loss of faith in them by the public.

Erosion of trust in expertise

Experts have always been under scrutiny and subjected to public disappointment if technology fails. Take the following extract from a book on the sinking of the *Titanic* in 1912.

> Overriding everything else the Titanic also marked the end of a general feeling of confidence. Until then men felt they had found the answer to a steady orderly civilized life . . . for 100 years technology had steadily improved. For 100 years the benefits of peace and industry seemed to be filtering satisfactorily through society. The Titanic woke them up. Never again would they be quite so sure of themselves. In technology especially the disaster was a terrible blow. There was the unsinkable ship — perhaps man's greatest engineering achievement — going down the first time it sailed . . . if this supreme achievement was so terribly fragile what about everything else . . . if it was a lesson, it worked — people have never been sure of anything since. [Lord, 1975, 38]

The loss of the *Titanic* was a worldwide news story. The press drew attention not only to the failure of the technology, but also to the role of the *California* in not responding to distress signals and the woefully inadequate provision of lifeboats. The inevitable inquiry sought to apportion responsibility and blame — an often explicit feature of inquiries since.

The Challenger Shuttle disaster has had a similar devasting effect on the American public. What began as a national tragedy for the loss of life became a national outrage over possible oversight and expediency. Again the catalogue of possible neglect, incompetence or corruption undermines the confidence in the managing institutions and the capacity of the regulators to guard against or police mistakes.

Evidence from a MORI poll conducted in 1985 of a national sample of British citizens indicated that few of those questioned, only 28 per cent out of 1,824, trusted scientists involved with major commercial companies to tell the truth about the impact of science and technology; only 12 per cent said they trusted civil servants, 11 per cent government ministers or their member of Parliament. Interestingly 31 per cent claimed to trust environmental groups and 38 per cent consumer goods. Most trust was vested in doctors (67 per cent), clergymen (57 per cent) or judges (46 per cent). American national opinion poll data quoted by Marrett (1984, 314) suggests that government and Congress similarly did not inspire much confidence from American citizens although in 1980, 41 per cent of a national sample indicated they had confidence in the 'scientific community'.

The model of safety regulation in Great Britain is one in which personal responsibility is deferred to the appropriate institutions. In other words the public expect to be protected by those institutions who are explicitly charged with the nation's health and safety. Scandals and accidents erode the confidence and trust that enables this system to operate effectively. With the failure of these institutions, the public (whether they be the locally affected or the general public) look towards other organisations to act as watchdogs on their behalf or who themselves become involved in local participation. In the United Kingdom this greater involvement marks the evolution of a political dimension to environmental threats.

Politicisation of risk

Active concern for the environment goes back to the nineteenth century (Lowe and Goyder, 1983). As a reaction against the *laissez-faire* attitudes of Victorian industrialisation, an intellectual tradition grew up questioning the costs of technology and lamenting the damage to flora and fauna. Early organisations such as the Commons Open Spaces and Footpaths Preservation Society and the National Trust were concerned with heritage and cultural preservation. The new wave of environmental groups born in the 1960s and 70s were quite different in composition and objectives. Stringer (1982, 46) argues that during the 1960s the alienated and underprivileged began to demand information, in advance, on planning decisions together with the right to comment. In the United Kingdom the principle of citizen participation was embodied in the 1968 Town and Country Planning Act. The formation of Friends of the Earth (UK) (FoE) in 1970 from the American original FOE spawned a range of environmental pressure groups such as Socialist Environment and Resources Association and Farming and Wildlife Advisory Group. These participate in inquiries, publishing well-argued information and orchestrating events sometimes outside the law.

O'Riordan (1986) contends that the issues are more than conservation or preservation of the environment, that the control of hazardous technology has become the focus for political dissent. 'Risk acceptance is rooted in the political culture, in views about fairness and justice, and in feelings about exploitation. If a person or a section of society thinks he/she is being unjustly treated then presumed dangers emanating from an unrespected agency will be despised'. O'Riordan (1983) previously argued that safety had become a focus for a growing community of counter-establishment scientists who do not have traditional institutional affiliations and who espouse alternative views about technological developments. Indeed much of the dispute over particular technological developments is focused on the counter-interpretations of the data (often numerical) collected for the purposes of risk assessment. Funtowicz and Ravetz in their chapter on managing the uncertainties of statistical information undertake a forensic analysis of the collection and processing of data. Their challenge to the policymakers is the status that is all too often assigned to the unassailability of 'the facts', frequently numbers, as inputs into decision-making. Wynne, in his second contribution, Chapter 8, takes this notion even further and not only gives alternative interpretations to the results of the quantitative analyses, but also questions the rationale for the use of decision analytical techniques such as cost benefit analysis.

Cumulative effects of social trends

Accidents, the role of the media and politicisation of environmental issues may imply that the developers of technology have been remiss in evolving systems and procedures to ensure safe operation of process or plant. This is not the case. Engineering sciences have established elaborate methods of risk assessment, probability analysis and fault tree approaches which involves the identification of possible sources of accidents and a calculation of their likelihood of occurrence. Alongside reliability engineering to eliminate potential weakness in design and the addition of fail-safe features, legislation and regulation add another layer of safety. Standards of good practice, limiting emissions to the environment and a reporting system for accidents are often mandatory. In addition, pressures for public accountability and involvement have led to the increasing requirement for environmental impact statements from developers, outlining in advance the possible threats to the environment and the local population. When the Nixon administration put through the National Environmental Policy Act (NEPA), it declared as its major goal the encouragement of a productive and enjoyable harmony between man and his environment, and required environmental impact statements for major developments. The Seveso disaster gave the impetus to similar requirements to be enacted in the European context. The Seveso Directive, issued by the EEC, places a general duty on manufacturers using a wide range of processes to prevent major disasters, limit any disastrous consequences, produce safety reports, provide on- and offsite emergency plans and inform members of the public about appropriate behaviour in the event of an accident.

However the technical risk assessments and legislative requirements have not been unequivocally successful in protecting the public from technological accidents and mishaps, although in certain areas there have been demonstrable successes. For example, in the United Kingdom smoke emissions from all sources were reduced from 0.37 million tonnes in 1976 to 0.27 million tonnes by 1986 (Central Office of Information, 1988). Sulphur dioxide emissions were reduced from 4.99 to 3.74 million tonnes in the same period. Yet there has been a rise in the number of complaints made to the Environmental Health Officers under the 1974 Control of Pollution Act and the number of pollution prosecutions increased during the corresponding period.

Despite improvements (or because of them) public awareness of the seriousness of threats to the environment or to people has increased. Social and Community Planning Research have run an annual survey of British social attitudes for the last five years. In the latest survey (Jowell *et al.*, 1987), when asked which of a series of environmental hazards represented a very or quite serious problem most respondents (over 90 per cent) cited industrial waste in rivers and the sea, nuclear waste from power stations and industrial fumes. In all cases the percentage of respondents endorsing these has progressively increased during the years that the surveys have been running. This burgeoning concern over environmental hazards and their management needs to be explained. Most of the chapters in the book take retrospective case study material and reconstruct the differing perspectives held by the various participants. One case study, that of biotechnology, is presented as an anticipatory example. Ravetz argues for the development of more imaginative and flexible mechanisms to undertake the scientific assessment and regulatory control of releasing genetically engineered micro-organisms to the environment. Brown discusses the emergence of public concern about the impact of biotechnology. Both draw from the experience of the nuclear industry to apply proactively to biotechnology.

Approaches

This book hopes to make its contribution towards reducing the present and future difficulties and tensions between those who develop, manage and bear the risks (as well as consume the benefits) of technological progress. The various contributors seek to do this by analysing risk controversies from a social science perspective. There is no unified social science theory of risk, rather there are a number of approaches and methodological tools that are used to identify the social processes underpinning definitions of, responses to and ways of managing technologies. These are neither mutually exclusive, nor maintain strict disciplinary boundaries, rather they are overlapping and mutually informing. These approaches may usefully be considered in terms of the level of analysis, i.e. institutions, communities, groups and individuals.

Institutional analysis

Fundamentally, management of risks, especially technological ones, involves people. They have to be warned or protected against adverse consequences and above all to be persuaded of the efficacy of scientific analyses and technological solutions to minimize threats to their personal safety or property or to society and the environment as a whole. There is a delicate balance between the public and risk managers based on trust and confidence. In an increasingly differentiated and complex world people do not wish to be unduly burdened or worried by the many possible risks in the environment; hence they institutionalise matters of safety to regulatory and policy-making bodies. This determent of personal responsibility can only work if the managing institutions are seen to be in control and have the appropriate knowledge to undertake this task on behalf of society. Social science analyses explore the relationships within and between the risk creators and risk bearers. The assumptions of scientific expertise and its use in policy are exposed, and the reactions of communities under threat examined.

Short (1984) sees the role of social science analyses as demystifying not only the unfamiliar, strange or esoteric aspect of risk management but also as questioning the familiar and 'taken-for-granted'. However on a cautionary note he writes (176) that 'disabusing ourselves and others of the ultimate solvability of such problems is an important aspect of the demystification function of the social sciences and it may be one of the greatest services we can perform for the larger society'.

O'Riordan (1982) declares that evaluation and control of technology are totally emeshed in political judgement. He proposes (99) that 'risk management is a political scientific process that is constantly adapting as scientific understanding improves as the political priorities alter and as the public mood modulates'. He further argues that the combined societal pressures from the environmental movement, consumer groups and civil rights activists who challenge secretive and autocratic decision-making have broadened the base for environmental risk assessment beyond the technical domain. Thus the idea that technology can control itself through its own methodologies and expertise becomes untenable.

An insightful account of the ways in which values and self-interest are inextricably woven into development, management and decision-making about technology is provided by Collingridge (1980). He demonstrates that commissioning or managing institutions get hopelessly locked into a particular cause by the very nature of their social structures that have little to do with an evaluation of the costs or benefits of the technology *per se*. Decisions to develop a particular technology involves a commitment of resources and vested interest for a successful outcome. Failure is unthinkable. Thus the institution can engage in a series of tactics to erode criticism or avoid discovery of mistakes.

Wynne (1982), argues quite forcibly that decision-makers create and foster ideas about an idealised notion of scentific method. This is because they need to justify the dominant role of professional expertise in exerting authority and control over technology. The public is also encouraged to assimilate this idealised view. If this is damaged then the contract of trust and confidence with the public is broken. Wynne's central proposition is that often mistakes are not individual lapses at all but are actually encouraged by particular institutional arrangements. The role of the expert and the relationship between expert and the public is discussed by Wynne and also by Thompson and Irwin in this volume. They seek to examine the status of expertise and the factors influencing the expert's problem definition and preferred solutions.

Community responses

Exposure to environmental threats at the community level has invoked the use of the explanatory variable of 'neighbouring'. This is defined by Unger and Wandersman (1985, 141) as 'the social interaction, the symbolic interaction and the attachment of individuals with the people living around them and in the place in which they live'. Neighbouring is rendered in terms of personal and emotional currency, i.e. sociability and instrumental support, i.e. helping. When the physical environment is threatening, the glue that binds the macro physical and social aspects of neighbourhood and its organisation becomes unstuck. Edelstein and Wandersman (1987) note a number of typical community responses in situations of threat they call environmental turbulence. First people resort to their known and existing social networks; second they turn to the institutional networks such as government agencies that are supposed to help them. Often the intensity of the disruption and failure of the institutional resources means that people form new networks and organise protest groups to counteract the adverse consequences. Environmental turbulence is characterised by feelings of loss of control, powerlessness and uncertainty. There is a need to restore some stability often by establishing action groups. The external threat can sometimes provide a new and/or stronger sense of identity for the community. They conclude (92) 'in summary one outcome of the toxic exposure appears to be the development of new sources of social support particularly where an area is isolated because of the shared exposure which accompanies and influences the development of a sense of community and the participation in a grass roots effort to address the problem'. However, they note possible negative outcomes with communities being divided and factions forming questioning the legitimacy of compensation claimants or disagreeing about the goals of their protest or actions to achieve their ends. Such varied community responses are evident from Macgill's study of reactions to the radioactive discharges explored in Chapter four.

Risk perception

Differences in perception have been the focus of much of the research effort by psychological risk analysts. Hitherto engineering and actuarial studies of the risk of technological facilities concentrated on their physical and measurable properties. It was recognised that discrepancies existed between such 'objective' indices and the 'subjective' public judgements. Much of the early innovative work was conducted by Slovic and his colleagues at the Oregon-based perceptronics group. Risk perception studies are concerned with understanding differential judgements of the riskiness of various hazards. Fischhoff *et al*, (1978), asked people to judge some 90 hazards on different criteria. Factor analysis showed three underlying dimensions: 'dread' risks comprising hazards that are judged to be fatal or catastrophic compared to those with less severe consequences; imposed, involuntary technological risks versus voluntary, known technical risks; and a societal/personal consequences factor.

Slovic and his group have also highlighted the importance of biases in perception. They propose that lay people not only use qualitative criteria by which to judge hazards but also are influenced by memorable instances. So a disaster graphically and dramatically portrayed, especially if it shows vulnerable individuals such as children at risk, may lead to erroneous beliefs about the likelihood of accidents. Slovic *et al* (1982) illustrate this process with reference to nuclear power, showing it as being particularly susceptible to perceptual biases. Differences are also observable within the scientific community. Lichter and Rothman (1983) show that scientists qualified in energy related fields had different attitudes towards nuclear power than those from other disciplines.

There are different kinds of judgements invoked to evaluate risk. Research summarised by Lee (1983) showed that there are differences, e.g. in the assessment made by individuals if they perceive the risk as personal to themselves or affecting society as a whole. According to Lee in the former case self-interest plays a predominant role in the assessment whereas in the latter moral values are more influential as the impact on the individual is more remote. Empirical research by Green and Brown (1985) support this conceptual differention. In other words, a perceived threat to self and a perceived threat to society are independent.

Other influences on perception include catastrophic potential, i.e. people's apparent greater reaction to a large number of deaths or injuries concentrated in one accident (say, a multiple car crash) than to an equivalent number of fatalities happening over an extended period of time. Similarly if the consequences of the catastrophe are delayed there is less concern than if they are immediate. Again differential judgements appear to be made if the risky process or activity is seen to be voluntary or involuntary. An explanation for this can be found in attribution theory. This formulation is concerned with the attribution of blame or responsibility. In the case of technology, as this becomes more complex and remote, people tend to see themselves as involuntary victims and hence lay blame on the controlling management rather than on themselves.

Theories of attitude developed by Fishbein and Ajzen (1975) provide important insights into psychological effects of environmental threat. Their expectancy-value model of attitude has been widely used in the study of public reactions to nuclear power. The basic assumption holds that there are both positive or beneficial and negative or adverse attributes of an object. The greater the belief in positive attributes, the more likely it is that a person will be favourably inclined towards that subject. Many studies using this formulation adopt a factor analytic method to reveal the cognitive structure of beliefs (Hughey *et al*, 1985; Otway *et al*, 1978) which shows that people make judgements about nuclear energy in terms of economic factors, environmental consequences, socio-political implications, psychological risks and local impacts. As one might expect, public attitudes become progressively more 'anti' as the possibility increases of a nuclear power station being built in their own locality (Van der Pligt *et al*, 1985). Furthermore, those who are in favour as compared with those who are opposed to nuclear power show marked differences in the importance that they attach to socio-economic, environmental and health consequences of building a nuclear power station. This fits with the research showing that attitudes are important mediators in the response to environmental stressors.

In summary, approaches to the analysis of environmental threat employed by social scientists attempt to find explanations for the differing perspectives held by the various social actors. Actors are defined as members of the general public, specific communities hosting a potentially hazardous facility, the legislators and regulators, scientists and technical experts. The factors that influence the way they identify risks to be concerned about, their assessment of the dangers, their beliefs about the possibility of controlling and managing plant or processes are revealed as relating to their past experiences, present goals and purposes, and future expectations. Differences given between them can be seen to derive from these social and psychological influences on the one hand, and the cultural and political reality in which they are embedded on the other. The case study examples referred to in the subsequent chapters provide powerful expositions of these explanatory concepts.

Methods

Given the development of society's concern about protecting those who work in potentially hazardous industries and the need to safeguard the surrounding population and environment, a series of techniques have emerged subsumed under the rubric of risk assessment. Early methodologies were concerned with physical safety and the focus was on minimizing deaths. Later this was extended to health effects and damage to the environment and most recently has incorporated social and political consequences.

Renn (1985, 112) provides a useful delineation of risk analysis as:

(1) The engineering approach which is concerned with quantification of risk levels and based on technical considerations.
(2) Decision analysis techniques which try and incorporate other features such as environmental damage and public concern as well as technical aspects.
(3) Risk perception studies which try and determine the different valuation placed on different risks by various groups in society.
(4) Policy analyses which attempt to explain the social and political influences on design and implementation of policy.

Technical risk assessment

The assessment of the riskiness of an industrial process or plant is largely the preserve of the engineering sciences. Green (1982) draws together a number of contributors describing these in some detail. Methods include the measurement of design safety criteria by failure mode and effect analysis, event tree analysis. These seek to identify possible accident sequences and calculate their probability of occurrence. The objective is to design out possible faults and anticipate failures in the system.

The stages in technical risk assessment include:

(1) identification of the source of the hazard and its causes;
(2) analysis of the mechanisms by which hazards could occur;
(3) analysis of the consequences of hazard;
(4) calculation of the likelihood of the hazard occurring;
(5) judgement about the acceptability of risk;
(6) implementing decisions.

In addition actuarial methods are used in which past statistical evidence of failures are calculated and an estimation of the expected rate of accidents calculated. Tables of comparative risk are presented with the calculated level of mortality against each hazard. These calculations often include the assignment of monetary value to the threat of mortality. The underlying logic of both technical risk assessment and the temptation to assign monetary values to human life is explored in detail by Thompson in Chapter six.

More generally, critiques of the technical or so-called 'objective risk assessment' have been offered by Lee (1986a). He characterises the rationale for having some objectively determined level of risk so that hazards may be placed on some unidimensional scale running from acceptable to unacceptable. The shortcomings are enumerated as follows: there is a dearth of dependable past experience for some kinds of risks, e.g. biotechnologies; quantification of the monetary value of life or the weighting of morbidity or other social costs of accidents is difficult and problematic; assessment of consequences may be influenced by who is affected, e.g. the differences between exposure of workers versus the public, or specifically children at risk; expert judgements are not uniform or consensual but are subject to occupational loyalties or institutional affiliation.

Decision analysis

Analytical techniques have developed to aid policymakers in formulating decisions about potentially hazardous technology. They are described by Renn (1986) as comprising ten steps:

(1) recognition that a decision problem exists;
(2) structuring the problem and specifying the goals or needs for the proposed action;
(3) choice of appropriate criteria or dimensions which can be used as a classification to assess consequences for each option;
(4) definition of alternatives or options that are technically feasible and will correspond with the aims specified at step 1;
(5) assessment of outcomes for each decision option according to the criteria list;
(6) assessment of probabilities for each outcome;
(7) assessment of weightings for each criterion;
(8) transformation of outcomes into utilities according to the client's perceptions;
(9) evaluation of alternatives by summarising the expected utilities of each possible outcome weighted by their perceived probability of occurrence;
(10) sensitivity analysis to investigate the effect of alterations with regard to the criteria composition, outcome generated and probability and preference assessments.

Underlying these steps are a series of assumptions which include beliefs about the negative outcomes of a particular event or activity. Risk only has meaning if it is regarded as relative to the values of the decision-makers. The aim is not an objective measurement of the physical properties of risk but a more consistent and logically sound model to represent the knowledge and preferences of individuals or institutions. Covello (1987) provides an overview of decision analysis and its use in risk management and gives a useful comparison with cost benefit analysis. He points out that both seek to rationalise and logically choose from alternative policies when outcomes are uncertain; both attempt to describe and quantify the relative advantages and disadvantages of options; both decompose decision problems into more manageable components; both try and integrate and synthesise information into a single index reflecting the value of the proposed course of action. Differences between them relate to their treatment of utility.

The concept of utility is basic to these approaches: the value or worth or desirability of the consequences of particular courses of action. A utility function is a scale of subjective values assigned by the decision-maker to the consequences of alternative decisions. The utility function captures the decision-maker's willingness to take chances and can express a risk taking, risk neutrality or risk aversive attitude.

Cost benefit analysis (CBA) tends to seek one benefit — maximising economic efficiency — while decision analysis (DA) incorporates several of many objectives including public concern. Similarly CBA typically only considers consequences in monetary terms, DA takes on board other consequences such as detriment to the environment or erosion of trust in experts. CBA often expresses consequences in terms of people's willingness to pay while DA tends to include all possible consequences of policy according to a weighted scheme deriving from the decision-makers' values.

Wynne (Chapter three) presents a critique of such methods in which he questions some of the basic assumptions underlying them and their omission of the social context in which risks are constructed. As has been shown in a previous section, experts and decision-makers are not immune from error or bias and these can be reflected in their analyses.

Perception of risk studies

The study of perception is concerned with the ways in which individuals learn about their environment through their senses. People have the capacity to store and organise the information they take in and these capacities influence the absorption of further information. Past experiences and individual capability do influence the selection, processing and assimilation of new information. An impetus towards studying perceptions of risk was the observation that the public seemed to have a different response from those of the expert to the apparent riskiness of technology. Methods used to measure both public and specialists' evaluations concerning risk have included social surveys and psychometric scaling techniques.

Surveys are composed of a series of questions either completed by the respondents themselves as a paper and pencil task or used to elicit verbal responses in a face-to-face interview. Problems with surveys lie with the framing of questions and the interpretation of answers. Brown and White (1987) show that the wording of questions about nuclear energy can have a significant effect on the proportions responding in a favourable or unfavourable direction towards the technology. In addition, surveys may not have appropriately controlled sample populations, which lead to possible biases in reported results.

Psychometric scaling techniques are used to enable people to rank potential hazards in order of riskiness or to evaluate hazards against a series of judgemental criteria. Typically this methodology uses factor analysis to interpret the data generated by the scales. Fischhoff *et al*, (1978) employed such techniques in their study. They identify three important features in people's judgement and some 90 hazards. These are 'dread' factors, voluntariness and exposure and scale of consequences. These and other studies have revealed important differences in the ranking of riskiness of hazards between the public and various expert groups. The technique requires people to make somewhat arbitrary comparisons, i.e. ones they are not likely to make in real life, such as the relative riskiness involved in skiing as compared to using pesticides. There are other criticisms of the factor analytic method, especially the labelling of dimensions or structures by the investigators, i.e. imposing their interpretation on the collective pattern of the data. Wynne (Chapter 3) examines the meaning of the dread dimension of risk proposed by Slovic and his co-workers and provides an alternative analysis which argues that dread is a code for the nature of the relationship between the knowledgable expert and the supposedly ignorant public. Thus the 'rational' expert understanding nuclear physics perceives nuclear power as technically controllable. The 'irrational' public, unschooled in reactor physics, resort to dread. Wynne suggests that the public are not being irrational but that dread signals the social distance between the expert and the public in their judgements about nuclear power.

In sum, methods or tools to analyse the potential consequences of risk have been derived to deal with technical, economic, social, psychological and political impacts. Within itself, any particular set of tools can be and is criticised in order to refine its effectiveness. The thesis of this book is not that technical risk assessment or economic analyses are wrong or should be abandoned, rather that they are incomplete and should be supplemented by features that are perhaps less obvious and more amenable to strictly quantitative analyses. The succeeding chapters draw attention to these qualitative features.

References

Brown, J. and White, H., 1987, 'The Public's Understanding of Radiation and Nuclear Waste', *Journal of the Society for Radiological Protection*, **7**, 61–70.

Canter, D., Craik, K., and Brown, J., 1985, Editorial: 'Psychological Aspects of Environmental Risk', *Journal of Environmental Psychology*, **5**, 1–4.

Central Statistical Office, 1988, *Social Trends*, **18**, HMSO, London.

Collingridge, D., 1980, *The Social Control of Technology*, Open University Press, Milton Keynes.

Covello, V.T., 1987, 'Decision Analysis and Risk Management Decisionmaking; Issues and Methods', *Risk Analysis*, **7**, 131–9.

Cvetkovich, G. and Earle, T.C., 1985, 'Classifying Hazardous Events', *Journal of Environmental Psychology*, **5**, 5–36.

Douglas, M. and Wildavsky, A., 1982, 'How can we Know the Risks we Face? Why Risk Selection is a Social Process', *Risk Analysis*, **2**, 49–51.

Edelstein, N.R. and Wandersman, A., 1987, 'Community Dynamics in Coping with Toxic Contaminants', in Altman, I., and Wandersman, A. (eds), *Neighbourhood and Community Environment*, Plenum, New York, 69–112.

Fishbein, M. and Ajzen, I., 1975, *Belief, Attitude, Intention and Behaviour: an Introduction to Theory and Research*, Addison-Wesley, Reading.

Fischhoff, B., Slovic, P., Lichenstein, S., Read, S. and Coombs, B., 1978, 'How Safe is Safe Enough', *Policy Sciences*, **8**, 127–52.

Green, A.E. (ed.), 1982, *High Society Technology*, Wiley, Chichester.

Green, C.H. and Brown, R.A., 1985, 'Injury or Death by Fire: How People Rate their Chances', *Fire*, March, 44–8.

Hughey, J.B., Sundstrom, E. and Lounsbury, J.W., 1985, 'Attitudes towards Nuclear Power; Longitudinal Analysis of Expectancy Value Models', *Basic and Applied Social Psychology*, **6**, 75–91.

Jowell, R., Witherspoon, S. and Brook, L., 1987, *British Social Attitudes; the 1986 Report*, Gower/Social and Community Planning Research, Aldershot.

Kaspersen, R. and Pijawka, K.D., 1985, *Societal Response to Hazards and Major*

Hazard Events, Clark University Hazard Assessment Group Center for Technology Environment and Development, Worcester, Mass (CENTED reprint 45).

Kates, R.W., 1981, Risk Assessment in Environmental Hazard; Scope Report 8. Wiley, Chichester.

Lagadec, P., 1982, *Major Technological Risk; An Assessment of Industrial Disaster*, Pergamon, Oxford.

Lazarus, R.S. and Cohen, S., 1977, 'Environmental Stress', in Wohlwill, J. and Altman, I. (eds.), *Human Behaviour and Environment*, Plenum, New York, 89–127.

Lee, T.R., 1983, *The Perception of Risks in Risk Assessment; a Royal Society Study Group Report*, The Royal Society, London, 94–148.

Lee, T.R., 1986a, 'Public Attitudes towards Chemical Hazards', *Science of the Total Environment*, **51**, 125–47.

Lee, T.R., 1986b, 'Effective Communication of Information about Chemical Hazards', *Science of the Total Environment*, **51**, 149–83.

Lichter, S.R. and Rothman, S., 1983, 'Scientists Attitudes Towards Nuclear Energy', *Nature*, **305**, 91–3.

Lord, W. 1976. *A Night to Remember*. Allen Lane/Penguin, London.

Lowe, P. and Goyder, J., 1983, *Environmental Groups in Politics*, George Allen and Unwin, London.

Market Opinion Research International (MORI), 1985, *Public Attitudes to New Technology*, MORI, London.

Marrett, C.B., 1984, 'Public Concerns about Nuclear Power and Science in Freudenburg', W.R. and Rosa, E.A. (eds), *Public Reaction to Nuclear Power; Are There Critical Masses*. Westview, Boulder, 307–28.

Mazur, A., 1984, 'Media Influences on Public Attitudes towards Nuclear Power in Freudenburg', W.R. and Rosa, E.A. (eds), *Public Reaction to Nuclear Power; Are There Critical Masses*, Westview, Boulder, 97–114.

O'Riordan, T., 1982, 'Risk Perception Studies and Policy Priorities', *Risk Analysis*, **2**, 95–100.

O'Riordan, T., 1983, 'The Cognitive and Political Dimensions of Risk Analysis', *Journal of Environmental Psychology*, **3**, 345–54.

O'Riordan, T., 1986, 'Politics in Risk and Risks in Politics', Paper presented to the First International Conference on Risk Assessment of Chemical and Nuclear Materials, University of Surrey, Guildford, 22–26 February.

Otway, H.J., Maurer, D. and Thomas, K., 1978, 'Nuclear Power; the Question of Public Acceptance', *Futures*, **10**, 109–18.

Renn, O., 1985, 'Risk Analysis; Scope and Limitations', in Otway, H. and Pelto, M. (eds), *Regulating Industrial Risks*, Butterworths, London, 111–27.

Renn, O., 1986, 'The Concept of Decision Analysis for Risk Assessment', Paper presented to the 1st International Conference on Risk Assessment of Chemicals and Nuclear Materials. University of Surrey, Guildford.

Short, J., 1984, 'The Social Fabric at Risk; Towards the Social Transformation of Risk Analysis', *American Sociological Review*, **49**, 711–25.

Sjoborg, L., 1987, (ed.), *Risk and Society; Studies of Risk Generation and Reactions to Risk*, Allen and Unwin, London.

Slovic, P., Fischhoff, B. and Lichtenstein, S., 1982, 'Psychological Aspects of Risk Perception', in Sills, D., Wolf, C.P. and Shelanski, V. (eds.), *Accident at Three Mile Island; the Human Dimensions*, Westview, Boulder.

Stringer, P., 1982, 'Towards a Participatory Psychology' in Stringer, P. (ed.), *Confronting Social Issues; Applications of Social Psychology*, Volume II, Academic Press, London.

Unger, D.G. and Wandersman, A., 1985, 'The Importance of Neighbours; the Social Cognitive and Affective Components of Neighbouring', *American Journal of Community Psychology*, **13**, 139–69.

Van der Plight, J., Eiser, R.J. and Spears, R., 1985, 'Social Judgement in the Field; Attitudes Towards Nuclear Energy', in Denmark, F.L. (ed.), *Social Ecological Psychology and the Psychology of Women*, Elsevier, Amsterdam, 333–46.

Wynne, B., 1982, 'Institutional Mythologies and Dual Societies in the Management of Risk', in Kunneuther, H. and Ley, E.V. (eds.), *Proceedings of a Summer Study on Decision Processes and Institutional Aspects of Risk*, Springer-Verlag, Heidelberg, 127–43.

Chapter 2

Deciding about Risk: Expert Testimony and the Regulation of Hazard

Alan Irwin

The problems of managing technological risk span many different issues and many forms of hazard are discussed in this book. A recurrent theme is the tendency of official bodies to portray hazard issues as if they were essentially *technical* in character — in other words, to assume (at least in public statements) that the 'facts' of any risk assessment can in some way speak for themselves. The purpose of this chapter is to look more closely at the nature of technical evidence within hazard disputes and to consider the limitations, as well as the more positive contribution, of such testimony within public policy-making. One important issue here is that of technical disagreement over matters of risk and the consequent necessity to decide public policy against a background of technical as well as social uncertainty. It is to be hoped that a greater appreciation of the place of expert knowledge within hazard disputes will lead to a better mutual understanding between officially recognised 'experts' and those groups which claim to represent the 'public interest' (and which may themselves offer a substantial — if contrasting and at times even contradictory — form of expertise).

If further evidence was needed to support the argument that official bodies have largely defined hazard problems in narrowly technical terms, then specific reference can be made to the control of chemical carcinogens in Britain. In areas such as pharmaceutical safety, pesticide control or the regulation of food additives, a series of advisory committees presently operate whose membership is formally constituted on the basis of appropriate expertise (Irwin and Green, 1983). For workplace hazards a system of tripartite representation has been in operation since 1974, but for most other areas of toxic chemical control in Britain the policy assumption has been that those with particular forms of expertise — in this case, epidemiology, medical science or toxicology — are uniquely placed to assess questions of safety and toxicity. In further support of this expert-based regulatory philosophy, it has been argued that expert advisory committees are especially well positioned to look impartially and objectively at the facts of a hazard issue and unlikely to get misled or be easily persuaded by emotion or pressure-group 'propaganda'. Thus, the technical analysis of hazard is placed at the core of the risk assessment process — with other factors such as social needs, or the viewpoint of potential victims generally being treated in a much more informal and implicit fashion (barring the occasional high-profile public inquiry conducted in such areas as nuclear power). In addition, 'appropriate expertise' as defined here would exclude the social sciences or the wider range of academic disciplines which touch upon questions of risk (for example, psychology or moral philosophy).

Although this emphasis on technical expertise remains a powerful force within public policy-making, a number of critical analyses have highlighted the limitations on such expertise in this context. In a series of publications, for example, Collingridge has argued that the effect of technical expertise on issues of public policy such as hazard control has actually been the *heightening* rather than the resolution of conflict (see, for example, Collingridge and Reeve, 1986). Thus, once a specific issue becomes a matter of widespread technical debate, policy resolution becomes almost impossible. In broadly similar vein, a number of sociological investigations have analysed the phenomenon of controversy within the scientific community and considered the relationship between such disputes and the social and institutional context within which they occur (among the more accessible of these studies are Nelkin (1979) and Meehan (1984)).

Taken together, these critical studies suggest a need to evaluate carefully the role of technical experts within risk disputes. In particular, they suggest that we must be much more sensitive to the uncertainties, inherent limitations and even contradictions of 'risk science'. Furthermore, it will be argued here that to do otherwise than acknowledge these characteristics is to invite a public backlash when bland reassurances and glib generalisations prove misleading (as they surely must). Going further, it will be argued that it is necessary to view risk decisions as unavoidably *social* as well as technical in character, so that we need to be conscious of the institutional commitments and implicit values which shape and direct decision-making processes.

As a starting point for such a social and technical evaluation, two short case studies will now be considered. Each examines the part played by specific groups of experts within policy disputes: first of all concerning the effectiveness of compulsory seat belt legislation and, secondly, the safety of the pesticide 2,4,5-T. The 2,4,5-T case will also be briefly discussed in Brian Wynne's chapter. The intention now is to complement that discussion in terms of the specific thesis outlined here. It should also be stressed that the intention is not in any way to single out these debates as examples of 'shoddy science' (which would be grossly unfair to the participants involved). Instead, the objective is to establish some general features of these debates which can stimulate future debate and assist our understanding of other such cases.

Compulsory seat belt legislation — a case of risk compensation?

Judged from a safe historical distance, the development of compulsory seat belt legislation in Britain (which currently remains law) seems like a story of slow, steady change and little controversy. Seat belts as a technology go back to the first decade of this century (as used in aircraft) and have been available to car buyers in Britain and the United States since the 1950s as an optional extra. When in 1973 the question of making seat belt wear compulsory in Britain came before Parliament for the first time, all cars built since 1964 had already been fitted with seat belts as a matter of legal necessity. Furthermore, the technical consensus remained solid throughout the 1970s (apart from one or two idiosyncratic statements by members of Parliament) that higher rates of seat belt wear would undeniably reduce road casualties. Indeed, to this day there has been little serious dispute that — once in a collision — seat belts offer considerable protection to vehicle occupants. Seen in this uncontroversial and incremental light, it is perhaps unsurprising that the British Parliament voted in 1983 to commence compulsory belt wear for an experimental period and in 1986 to continue the legislation indefinitely.

What such a Whig history of seat belt legislation in Britain significantly omits, however, is any account of the genuine disagreements and technical uncertainties which became apparent during the early 1980s. In large part, these disagreements and uncertainties have been linked to the research findings and public interventions of one man — John Adams, a lecturer in Geography at University College, London. The nature and conduct of the ensuing debate reveals much about the constraints and pressures under which real world risk disputes are conducted. At the very least it becomes clear how far 'risk science' can be from the usual portrayal of apparently cool and objective 'laboratory science' with its strict and supposedly impartial rules of debate.

Put at its simplest, Adams's intervention represented a fundamental challenge to the *ceteris paribus* assumption implicit in previous research work on this topic; perhaps the best single summary of Adams's research is found in Adams (1985). Essentially, this assumption took it for granted that one could safely generalise from laboratory studies of anthropomorphic dummies and field investigations of accidents to the actual responses of human beings following compulsory seat belt legislation. According to this consensus view, if the percentage of seat-belted occupants could be increased, as the legislation undoubtedly achieved in 1983, then it must follow that the level of protection would rise by an equivalent amount and that many deaths and serious injuries would consequently be avoided. On this basis, it was authoritatively estimated in 1976 that such a law would save approximately one thousand lives per year and either prevent or lessen the severity of some tens of thousands of injuries. If this logic seems merely 'common sense' then it should also be observed that its development stems directly from the engineering approach to road traffic safety which has been successfully developed (especially in the United States) since the 1960s (this point is discussed in Irwin, 1985).

Instead, what Adams presented in 1979 onwards was an inversion of this 'common sense' in a manner which many dismissed as counter-intuitive and nonsensical. Undeniably, however, Adams's intervention reveals just how sensitive scientific argumentation can be to the starting assumptions upon which it is built — assumptions in this case about real world behaviour which, given their implicit modelling of human responses, seem as much social as technical in origin.

The starting point for Adams's analysis was a statistical comparison of the actual impact of compulsion on casualty figures in various countries. Such data on seat-belt effectiveness had not previously been made available and immediately caused a furore among technical experts involved in the debate (many of whom were already on public record as supporting compulsory legislation). In summary, Adams's data suggested that legislation in various countries had had no beneficial effect on casualties — and, indeed, that the greatest decline in casualties had occurred in those countries lacking such legislation. In order to explain these startling findings, Adams proposed the notion of 'risk compensation', i.e., that humans do not behave like laboratory models but will actually adjust their behaviour in relation to the perceived level of safety. According to this hypothesis, we all respond to greater feelings of security by taking more risks than we would otherwise.

> Rock climbers with safety ropes, trapeze artists and tightrope walkers with safety nets, steeple jacks or novice gymnasts with safety harnesses all attempt manoeuvres with their safety equipment that they would not attempt without it . . . [Adams, 1982, 2–3]

It is important to note that Adams was not contradicting the evidence concerning the benefits of seat belts once involved in a collision. His argument instead was that behavioural modification (in the form of a risk compensation) leads to more dangerous driving performance and, since seat belts can bestow neither immortality nor invulnerability, to greater numbers of casualties. Thus, Adams behavioural/statistical analysis was based on very different underlying assumptions to the currently dominant engineering approach.

Even on the basis of this brief synopsis, a number of important points emerge concerning the nature of technical disagreements. The first of these must be that despite the somewhat vitriolic public statements made by certain participants the emerging dispute does not seem to be a mere question of 'good' and 'bad' science. Had it been such, then the dispute could — at least in principle — have been easily resolved. Instead, rather than proving or disproving each other's accounts, a central feature of the controversy was the manner in which those on either side seemed to talk past rather than to each other, so that proof or disproof became impossible when there was no agreement on the 'rules of the scientific game'. Thus, the concept of 'risk compensation' appears inherently untestable except through the kind of statistical analysis which Adams alone seemed prepare to conduct. The scientific antagonists would point to the strength of their own evidence but show little interest in replicating the techniques of the other, apparently expecting the truth of their statements simply to shine through the clouds of uncertainty. Furthermore, Adams' testimony was unfalsifiable since risk compensations can occur at any point and at any level. Thomas Kuhn has used the term 'paradigm incommensurability' to refer to such situations, where groups of scientists are operating within mutually incompatible sets of assumptions about a problem area and therefore find constructive dialogue virtually impossible (Kuhn, 1970).

Taking this point somewhat further, it would be misleading here not to comment on the acrimony and apparent frustration on either side which characterised this debate in the early 1980s. While there is no evidence whatsoever to impugn the honesty or personal integrity of the major participants, accusations were certainly made on both sides that debate was being advanced by innuendo and *ad hominem* attack rather than by open and direct technical scrutiny. Adams, for example, has alleged that the usual channels of scientific communication were closed to him because of his 'heretical' views; he argues that the scientific journals were reluctant even to devote space to a presentation of his case. Those in the opposite camp have criticised Adams for his use of 'non-standard' methods of dissemination such as direct circulation to MPs. In particular, they assert that he should have discussed his findings more widely within the technical community before 'going public'. On both sides of the debate, rhetoric was extensively employed as a means of undermining the alternative viewpoint. The logical incompatibility of the two schools was thus reinforced by a breakdown in communication at a more social and institutional level.

Following on from this discussion of the actual conduct of the debate, the key policy point to be made here is that while such phenomena as incommensurability and communication breakdown are by no means unique to risk debates (as the literature on sociology of scientific knowledge so clearly demonstrates), the social and institutional pressures on scientists must be all the greater when questions of public welfare are being discussed. After all, each side in this dispute was genuinely concerned about the impact, whether positive or negative, of seat belts on road casualties. It must also be said that the perceived pressure to achieve closure in the debate, i.e. to resolve it one way or the other — and fast, may not be entirely conducive to the maintenance of serious dialogue.

The next point which must be underlined about 'risk science' in this context is the unavoidable requirement for all sides to 'scale up' their findings from very specific and controlled analysis to the real world — in other words to model the complexity of actual human activity on the basis of known data. As has already been suggested, this modelling cannot be restricted to technical assumptions, for example regarding the transmission of decelerative forces through the human body in the event of external collision. It must also take account of changing behaviour patterns and questions of individual psychology. It might be argued that this point applies solely to the seat-belt controversy because of its dependence on human compliance and the methodological uncertainties introduced by changing variables such as the character of the vehicle population and socio-economic factors. However, as the case of 2,4,5-T and numerous other examples suggest, for example, evaluation in the event of fire or explosion, all areas of risk assessment inevitably involve some modelling of the 'human variable', an activity on which natural scientists do not seem uniquely equipped to pass judgement.

The heterogeneity of risk science adds considerably to this complexity. As we have seen, where no single discipline holds authority (as seems unavoidable in a multi-disciplinary area such as risk assessment) the contestable nature of any modelling exercise must be further exacerbated. None of this should be read as a criticism of scientific method for somehow 'failing us'; at present, these difficulties seem inevitable. What the analysis so far does suggest, however, is that the leap from the carefully controlled world of the laboratory to the realities of hazard exposure begs many questions of a social as well as technical nature. The starting point for future progress must be the open acknowledgement of this reality.

Given all of these logical and social uncertainties in the assessment of compulsory seat-belt safety, the inevitable final question to ask is how any policy judgement could eventually be reached. It is not surprising from the analysis so far that technical consensus was not achieved (indeed it seems unachievable). Instead, closure of debate was essentially a matter of institutional forces taking effect so as to sweep aside uncertainty. Despite the inherent uncertainties, by the early 1980s a powerful array of institutions had given their support to compulsory seat-belt use as the most effective means of 'doing something' about road traffic accidents. There is not sufficient space here to review the full chain of events or the technical disagreements which predictably followed the innovation of compulsion on an 'experimental' basis (this is available in Irwin, 1987). Instead, it can simply be observed that the eventual decision reflected the prevailing balance of institutional power in favour of legislation, a balance which seemed to be far more a matter of political preference than of obvious technical consensus.

Taking together all of these points concerning the technical and social debate, the major dilemma which remains is that of how such crucial policy issues are to be resolved when complete technical consensus does not exist. Moreover, it has been suggested for at least this case that such consensus may in fact be unachievable given the uncertainties of risk science and the competing scientific worldviews which co-exist. Discussion so far has also suggested that judgements are likely to be as much social as technical in character. The technical dilemma is thus one of establishing a sufficient 'burden of proof' when little certainty exists. The equivalent, and overlapping, policy dilemma is of constructing appropriate institutional mechanisms for the evaluation and assessment of the social factors which underpin every stage in risk debates. The second brief case study is designed to consolidate these points and to explore further their significance for the whole social process of 'deciding about risk'.

2,4,5-T — a hazard beyond all reasonable doubt?

Many of the above arguments concerning the difficulties and uncertainties (both technical and social) of 'deciding about risk' are echoed in the long-running controversy over 2,4,5-T. Although this herbicide has been produced since the 1940s, it is most commonly associated with the 'Agent Orange' which was sprayed by American aeroplanes during the Vietnam War as a defoliant. Since that time, 2,4,5-T has been used in numerous weed-killers available to both industrial and domestic users. In Britain, it has been particularly employed in dealing with woody weeds of the bramble variety.

Wynne (Chapter 3) will argue the inadequacy of interpreting this dispute as a conflict between irrational workers and compromised scientists; instead it can be more satisfactorily portrayed as a confrontation between different conceptions of the realities of pesticide usage. In terms of the special concern of this chapter with the relationship between technical expertise and public policy-making, a number of other points about this controversy can also be made. In particular, the confrontation between a number of British trade unions (led by the farmworkers) and the relevant official advisory committee, the Advisory Committee on Pesticides (ACP), raises many questions about the most appropriate 'burden of proof' which should be applied in such cases. As will be seen, the apparently technical decision over when a risk becomes proven seems very much a matter of institutional judgement as to what constitutes 'sufficient proof' rather than of the 'facts' simply speaking for themselves. In the sometimes highly charged area of risk assessment, such questions have considerable potential to polarise different groups in terms of their policy demands.

2,4,5-T has been controversial for some time because of its allegedly hazardous properties. Linkages have been claimed between the pesticide and such effects as chloracne, birth defects, spontaneous abortion and cancer. In the face of such evidence, many nations have banned or severely restricted the use of 2,4,5-T: among them the United States, Canada and the Soviet Union. For the same reasons, an international campaign has been mounted to restrict usage of the pesticide in developing countries.

Given this international concern, it is not perhaps surprising that a number of British groups have campaigned for the banning or strict control of 2,4,5-T, a campaign which has had some success in that many local authorities have agreed to cease spraying as have several major users such as British Rail, the Water Authorities Board, the National Coal Board, and the Central Electricity Generating Board. Very significantly for the present discussion, however, the ACP has steadfastly refused to ban the chemical on the grounds that it 'offers no hazard' to users or the general environment 'provided that the product is used as directed'. As Brian Wynne will argue, this difference of opinion can partly be explained by the disparity between the tidy world of the laboratory and the 'real world' of pesticide usage. Equally important for an understanding of risk decision-making is the approach which the two sides in the debate took to the larger issues of assessing chemical safety. It is on this dimension of the issue that the present discussion will now focus.

In order to explore this question, brief reference is needed to the form of technical evidence put forward by the two main parties to the dispute (simplified here as the 'trade unions' and the ACP). If we take one crucial year in the controversy as our focus, 1980, we can examine the evidence put forward by the trade union group and the official response which it subsequently received.

In 1980, the National Union of Agricultural and Allied Workers (NUAAW, now merged with the Transport and General Workers' Union) presented its 'dossier' on 2,4,5-T to the Minister of Agriculture for the ACP's assessment. By that time, the question of the pesticide's safety had already been referred to the committee no less than eight times. The NUAAW dossier — entitled *Not One Minute Longer* (NUAAW, 1980) — is broad in its coverage. Individual sections deal with the 'realities' of pesticide usage, the alternatives, a critique of previous ACP reviews, and (very importantly) a number of case studies where health damage is portrayed as following exposure to 2,4,5-T.

These case studies, which essentially represented the new evidence for the ACP's consideration, were largely drawn from a questionnaire which the NUAAW had circulated to its members through its newspaper, *The Landworker*. Such a move by a trade union to collect its own evidence rather than rely upon external scientific assessment is interesting in itself as an attempt at formulating some form of 'counter-expertise'. The methodology employed by the NUAAW of going directly to its own membership also reflects a belief that farmworkers possessed a richer stock of experience than had so far been considered by the ACP.

Questions in the survey included the following: when did you last use 2,4,5-T? By what method? With what protective clothing? Medical information was also requested: have you suffered from the following symptoms after administering 2,4,5-T? Do you suffer from any of these medical conditions? In particular, the medical section requested information concerning difficulties in conception and pregnancy, birth deformities and genetic disorders.

In formal scientific terms, of course, this form of data collection was very much open to challenge for its self-selected sample and reliance on the testimony of victims rather than on the more officially recognised evidence of health professionals. For the union, it represented an attempt not necessarily to prove a causal link but to demonstrate evidence of sufficient doubt against the ACP's previous appraisals as to merit a ban on the pesticide's usage in Britain. The methodology chosen was, therefore, heavily influenced by this overall political objective.

The case studies as presented in the NUAAW dossier (altogether involving some 14 named individuals) are — as the union concedes — brief and follow a highly succinct format. Once again, this would be defended on the grounds of insufficient resources for a full survey, and also in terms of the overall objective, i.e. to cast doubt rather than make authoritative cause and effect statements. Thus, one 'victim' is described as having had . . . 'a miscarriage in 1977 and later the same year gave birth to a daughter . . . who has a cleft palate and a hare lip. Her husband . . . had been using

2,4,5-T when he worked for the Forestry Commission' (NUAAW, 1980).

This and similar information was then presented to the ACP in summary form accompanied by a more general review of the evidence so far. The overall conclusion of the dossier was:

Considering the additional evidence which has not been evaluated by the ACP, the existence of alternative weed killers and the overall lack of information about the effects on users of 2,4,5-T . . . it becomes absolutely incomprehensible that workers, their families and the general public can remain subject to the risks for one minute longer. [NUAAW, op. cit., 20]

The basis of the farmworkers' case to the ACP was that while their new evidence might in itself have been methodologically limited and the overall state of knowledge less than complete sufficiently uncertainty existed to justify a ban on the substance. Indeed, the argument was that it is precisely because of the level of uncertainty that a ban was justified. The sharp contrast between this social treatment of uncertainty and that of the ACP will shortly become very apparent.

The advisory committee's published response appeared later in 1980 as the *Further Review of the Safety for Use in the UK of the Herbicide 2,4,5-T* (Advisory Committee on Pesticides, December 1980). This review is considerably longer than the NUAAW dossier and summarises all major scientific developments since the ACP's previous report. It presents a detailed appraisal of the evidence and includes a number of appendices on topics ranging from environmental effects and operator exposure to considerations of alternative pesticides. Even the most superficial comparison of the two documents, therefore, reveals something of the relative technical resources available to each group.

On the specific issue of the trade union dossier, the ACP devotes one section of its report to a consideration of the NUAAW's case studies, concluding for each case that insufficient evidence exists to correlate the medical condition with 2,4,5-T or at least that it seems highly improbable that such a correlation could exist. In the above-quoted instance of miscarriage and birth deformity for example, the employment records of the father in question were first of all checked. Following this, the two parents and also the family doctor were interviewed in order to establish the level of exposure involved and the scale of alleged effects. The specific conclusion eventually reached regarding this case by the ACP was that: 'The type of deformity occurring in this case is common genetically. Mrs K's only possible contact with 2,4,5-T was through handling her husband's working clothes; and the likelihood of her having absorbed sufficient to have produced any toxic effect is remote in the extreme'. (ACP, op. cit., 35)

In general conclusion, the ACP argued that 'there are no grounds to suggest a causal relationship with the stated effects'. The underlying argument is further elaborated in a discussion of miscarriages and birth deformities (which were the most common health defect linked to 2,4,5-T). The committee summarises its assessment of the evidence by asserting that these cases 'neither implicate nor absolve' the pesticide as a cause of miscarriage or birth deformity. What they do illustrate is that with so many such cases occurring every year:

It can be relatively difficult to establish a linkage with environmental agents like pesticides. The reality is that some women who have been in contact with such an agent are likely to miscarry, and that some are likely to bear malformed children; but this in itself does not add up to cause and effect. Indeed, statistically it would be remarkable if families in contact with particular products such as 2,4,5-T herbicides were spared from these misfortunes. [ACP, op. cit., 13]

The ACP was thus making a powerful scientific argument in refutation of the NUAAW dossier. In effect their treatment of uncertainty turned the union evidence totally on its head. The committee's case was that the linkage between 2,4,5-T and health disorder was insufficiently substantiated and, therefore, that no rational argument could be made for a ban. The union dossier was dismissed as representing no more than a collection of highly anecdotal — and certainly circumstantial — allegations. Until firmer evidence became available, the correct policy decision should be to continue usage of the pesticide.

The rationality of the ACP case may seem watertight. However, a closer examination reveals that what separated the two sides in this debate was not essentially a technical difference of opinion; indeed, the farmworkers would acknowledge that their dossier did not prove a causal link with 2,4,5-T (believing as they do that the whole issue has been under-researched). Underlying the dispute there seems rather to be a much more judgemental question of the best approach to be adopted in cases of apparent uncertainty and in particular towards the whole issue of burden of proof.

As one of the trade union participants expressed it subsequently, there are at least two levels of proof which can be applied in such cases (Cook and Kaufman, 1982, 69–70). First of all, 'beyond all reasonable doubt' as applied in criminal law whereby a chemical (or a defendant in a murder trial) is deemed innocent until proven guilty, i.e. until all other alibis, alternative causes and hyptheses have been found inadequate. Second, there is the 'balance of probabilities' concept as found in civil law where a reasonable suspicion has been raised about culpability and sound judgement suggests that a causal link is the most likely explanation. The trade union view was clearly based on the second level of proof; the ACP seems to have been geared to the former. In practice, therefore, each group took a very different attitude towards 'circumstantial' evidence, the unions believing that any significant level of doubt suggested the need for a ban. The advisory committee, on the other hand, demanded a much more unambiguous demonstration of 'guilt' before it could feel justified in taking any regulatory step. Another trade union participant has recently summarised the impasse:

Here a crucial difference between our two approaches emerges. 'We will rescind its clearance if the union can prove to us that 2,4,5-T is harmful', was in effect what the ACP told the union delegation. 'No', we responded, we cannot supply proof 'beyond all reasonable doubt'. Our yardstick is to estimate the hazard on the basis of what we know, and if 'on the balance of probabilities', the substance appears dangerous, then it should clearly be taken off the market. [Kaufman, 1986, 169]

This attitude towards what constitutes 'sufficient proof' inevitably has consequences also for the kinds of evidence each would consider as satisfactory. In addition to their acceptance of relatively circumstantial data, the trade unions were far more willing to accept the validity of animal data. Their argument would be that given the lack of human evidence animal data is sufficient to cast a substantial question mark over the pesticide. On the other hand, the ACP preferred to await more solid human evidence drawn from at least one fully controlled epidemiological study. Until such 'scientific' data becomes available how could any regulatory step possibly be justified? A similar emphasis on human evidence has been identified in other British advisory committees regarding cancer risks. 'Our argument was and still is, that we are not prepared to supply the sort of evidence that they seem to insist is necessary — namely a sufficiently large number of proven victims, a high enough body count . . . We will not have workers used as human guinea-pigs any more'. (Ibid.)

Again, the above quotation illustrates extremely vividly the potential for polarisation which is contained within contemporary risk disputes (as one group's scientific evidence becomes another's body count). The other main point to be made in the context of this discussion is that clearly the level of proof upon which one insists in such cases is very much a question of social as well as technical judgement. Put differently, attitudes towards the assessment of pesticide risk are likely to be coloured by more general attitudes to the agro-chemical business and the desirability of pesticide usage on the current scale. Institutions similarly are likely to draw the line of 'sufficient proof' at different points depending upon their underlying commitments and perceived interests. This is certainly not to devalue the contribution of technical evidence to policy-making but rather to emphasise that such evidence can only be interpreted once prior judgements have been made concerning the appropriate burden of proof to be applied. Such social judgements seem unavoidable, but it may be that a wider admission of their necessity would allow a more general discussion of the most appropriate institutional forms and 'regulatory philosophies' to be applied.

Just as with the case of compulsory seat-belt wear, therefore, we can see the inter-relationship between apparently technical decision-making and the institutional setting in which such decisions are made. One concrete demonstration of this point has come in the case of 2,4,5-T where the trade unions have broadened their demands from a call for a ban on a specific herbicide to an insistence on institutional changes in the whole decision-making structure. In particular, they argue on the basis of experience to date, that unless they have a direct input to the decision-making process, 'in appropriate' levels of proof will continue to be required throughout the whole area of pesticide safety.

'Deciding about risk' clearly requires the highest standard of technical evaluation in order to inform and substantiate debate. However, our two examples strongly suggest that social and institutional factors must also play an important part. The latter case, in particular, reveals the kind of alienation and resentment which can develop when dialogue between expert advisors and potential victims break down (or is never established). The farmworkers' conclusion is that their voice needs to be heard within this process rather than excluded on the grounds of the insufficient scientific nature of their evidence. Their strong argument is that they can have little influence over regulatory policy so long as the present 'burden of proof' continues to be applied. Thus the key to decision-making is not further technical argumentation but changes at an explicit institutional level so as to ensure their participation in all relevant policy matters. We therefore return to the question of how society should make decisions in this climate of both technical and social uncertainty. This question will form the basis of the final section of this chapter.

Conclusion

The theme of technical expertise and its relationship to the decision-making process over risk has raised a number of points which deserve further discussion. In particular, the case study of compulsory seat-belt legislation reveals something of the policy difficulties raised by technical disagreement, especially when such disagreements seem remarkably resistant to resolution for reasons which are at least partly linked to the framework of social assumptions within which different social groups operate. This case also indicates the more general difficulties and ambiguities of 'risk science', with otherwise arcane debates taking place under enormous political pressure to find an acceptable 'solution'. As we have seen, the actual conduct of such debates can be very different from the stereotypically calm and objective world of 'laboratory science'. Once again we can identify the effects of the social as well as explicitly technical framework within which policy decisions must be taken.

The case of 2,4,5-T as presented here reinforces many of the above points but also introduces at least two new issues. The first concerns the 'invasion' of what has hitherto been the territory of technical expertise by non-expert groups. Not only were the farmworkers dissatisfied with the conclusions of the ACP, they also took the additional step of formulating their own brand of 'counter-expertise' in order to engage the expert committee on its own previously inaccessible ground. The arguments which followed may have caused some discomfort to the committee by exposing its membership to an unusually direct form of criticism — both at the levels of technical argumentation and of demands for a more participatory structure. The NUAAW case could indeed be considered as lacking in methodological rigour, at least in terms of the more rigid criteria of evidence employed by the ACP. However, this phenomenon of 'counter-expertise' is likely to grow as a feature of risk debates as those in the real 'firing line' attempt to include their own (often harshly acquired) experience within the debate. Committees such as the ACP will come under increasing pressure to take such experientially based evidence as seriously as possible.

The worst outcome of this new presure would be for existing decision-making structures to be seen as inflexible in the face of the testimony put forward by those who will actually have to suffer the physical costs of hazard exposure. This point does, of course, link very closely to the other main issue in the 2,4,5-T case study, the social judgement over where to 'draw the line' in terms of hazard evidence. Once again, those who will be put at risk by any over-tolerant regulatory philosophy are likely to need (and deserve) the maximum reassurance that their point of view has been fully taken into account, and that the balance of uncertainty is tilted in their favour rather than that of industrial organisations.

Taken together, these arguments concerning the increased social and technical complexity of risk decisions raise a number of dilemmas for future policy-making. How are viable policies to be established within this context of explicit technical uncertainty and enhanced demands for the potential victims to have a more powerful voice? One route which no longer seems tenable is for policy bodies to retreat into bland reassurances that the best technical advice has been consulted. Such appeals to a spurious technical and social certainty are likely to prove counter-productive and lead to a discrediting of existing institutions. Instead, a more positive response seems necessary.

In particular, there is a need for moves towards greater institutional flexibility so that a wider range of voices can be listened to and — equally important — feel that they have a real participation in the decision-making process. It would be unfortunate, and potentially disastrous, if the institutional response to the kind of uncertainties addressed here were to retreat into a technocratic reverie. The opportunity should rather to be taken to build constructively and imaginatively upon social concern in order to establish new forms of creative discourse. The starting point for this must be an acknowledgement of the fine social judgements which are an essential part of risk decision-making. Official acknowledgement should at least enable all concerned to be aware of the agenda for discussion.

Such progressive steps as broadening the level of participation, encouraging a social and political debate over where the 'burden of proof' should fall and granting a platform to 'deviant' technical experts cannot be presented as a quick-acting panacea for all policy difficulties. Certainly, they would require the accordance of legitimate status to a whole series of critical voices (such as trade unions, consumer and environmentalist groups, the hazard centre movement). They would also raise a number of vexed questions concerning the comparative resources and political influence available to various groups in society.

For those scientists who are presently in a position of relative authority the experience might initially be somewhat disconcerting as implicit models of how to evaluate risks come under challenge and areas of technical uncertainty become apparent to all. It seems likely nevertheless that in the longer term this dialogue would lead to a far more satisfactory mutual understanding over the difficulties and anxieties currently faced by all groups (whether those like the farmworkers who feel themselves to be putting their lives at stake or advisory committee members who presently assume a position of deciding what is an acceptable risk for others to suffer). Put less equivocally, there seems no hope of resolving the kind of policy dilemma explored in this chapter without such a dialogue. It may be hoped that the arguments here and in other chapters of this book will add support to those who are currently attempting to open up for debate the social and technical basis of risk decision-making.

The main focus of this chapter has been on technical experts and their role in public policy-making. It has been argued that this role is likely to become even more demanding in future as various social and institutional changes take effect, not least a growing climate of questioning and criticism. The sheer technical complexities and uncertainties involved in 'deciding about risk' will also become more widely recognised so that statements from official bodies will need to be more qualified and explicit as to their underlying assumptions. All this represents a challenge to scientists and technologists which should not be underestimated — nor apparently can it be avoided. It is undeniable that a full technical appreciation of new and existing hazards will be more necessary than ever as part of the social dialogues envisaged here. The real scientific challenge is thus to respond positively to this emerging climate in order that policy can be widely adjudged to be both effective and equitable.

References

Adams, J.G.U., 1982, 'The efficiency of seat belt legislation', SAE Paper 820819, Warrendale, PA (also published in *SAE Transactions*, 1982, 2824–38).

Adams, John, 1985, *Risk and Freedom: the record of road safety regulation*, Transport Publishing Projects, London and Cardiff.

Advisory Committee on Pesticides, 1980, *Further Review of the Safety for Use in the UK of the Herbicide 2,4,5-T*, MAFF, London.

Collingridge, D. and Reeve, C., 1986, *Science Speaks to Power: The Role of Experts in Policymaking*, Frances Pinter, London.

Cook, J. and Kaufman, C., 1982, *Portrait of a Poison: The 2,4,5-T Story*, Pluto Press, London.

Irwin, A., 1985, *Risk and the Control of Technology*, Manchester University Press, Manchester.

——— 1987, 'Technical Expertise and Risk Conflict: An Institutional Study of the British Compulsory Seat Belt Debate', *Policy Sciences*, **20**, 339–64.

Irwin, A. and Green, K., 1983, 'The control of chemical carcinogens in Britain', *Policy and Politics*, **11** (4), 439–59.

Kaufman, C., 1986, '2,4,5-T: Britain out on a limb', in Goldsmith, E. and Hildyard, N., *Green Britain or Industrial Wasteland?*, Polity Press, Cambridge, 165–71.

Kuhn, T.S., 1970, *The Structure of Scientific Revolutions*, University of Chicago Press, Chicago.

Meehan, R.L., 1984, *The Atom and the Fault: Experts, Earthquakes, and Nuclear Power*, MIT Press, Cambridge, Mass. and London.

National Union of Agricultural and Allied Workers, 1980, *Not One Minute Longer! The 2,4,5-T Dossier*, NUAAW, London.

Nelkin, D. (ed.), 1979, *Controversy: Politics of Technical Decisions*, Sage, Beverly Hills and London.

Chapter 3

Frameworks of Rationality in Risk Management: Towards the Testing of Naïve Sociology

Brian Wynne

Introduction

The authority and effectiveness of government policies for regulations and managing hazardous activies often founder upon disagreements about risks. Expert assessments of potential harm (usually defined as deaths, sometimes including injury and other damage) often differ, but such differences are usually minor compared to the huge 'perception gap' evident between experts and public groups of one kind or another. Such crippling disagreements are normally attributed to ignorance or subjective bias which is said to distort public understanding of the objective expert knowledge underlying official stances. Remedial approaches therefore focus mainly on ways of better informing the public, thus correcting their 'misperceptions'. Research on public risk perception has been traditionally performed within this paradigm, that on the one hand are the real risks as analysed by the experts, and on the other are the unreal perceptions constructed by ill-informed and emotive publics.

There will always be room for fuller public understanding of technical matters. However focusing on this aspect alone obscures a more important factor in the expert framing of risk issues. Lack of recognition of this factor distorts and impoverishes the decision-making process, and encourages the very polarisation that the experts lament.

While a substantial analysis of the deeper rationality of public risk perception[1] has accumulated, there has been little analysis of the embedded social languages of scientific risk analysis. This is an attempt to sketch out how to fill that hole, and to explain why it is important to do so. In this chapter I argue that expert as well as lay risk perceptions are necessarily founded upon deeper social models and assumptions. These are usually inadvertent and implicit in technical analysis, yet they appear to be the main underlying source of disagreement about objective risks. Expert assumptions about the social world of risk practices are necessary in order even to frame a technical risk analysis. Yet their assumptions and commitments in this domain may be no better than those of the lay public's — indeed they may be worse.

I argue that however sophisticated and liberal it becomes, reiteration of the conventional 'objective versus perceived' language of risk conflicts is not only provocative but undermines the credibility of the institutions that deploy it. Instead I suggest that we need to develop better routine institutional arrangements for unearthing and examining the buried social assumptions that structure expert and lay argument about risks.

Framing the problems

Conflicts about risks have become a central problem in current policy management of technology and environmental pollution. The most striking example is nuclear energy, but so-called public misperceptions of risks and associated misunderstandings of science affect innovation and management in, among others, biotechnology, pesticides, drugs, food-additives, and hazardous wastes. The Royal Society followed up its interest in risk assessment with a *Report on the Public Understanding of Science*[2] in which it was asserted that public misunderstanding of science is the root cause of the general objective-perceived risk hiatus between experts and 'the' public. Public illiteracy in science and maths was also identified as the main reason for widespread anxiety about biotechnology at a major international seminar in 1987.[3] The depth of emotional anxiety which itself inspires the 'rational expert' perspective has been most vividly shown in the nuclear field, where public opposition has been equated to a threat to civilisation.[4] However nuclear power is no longer alone in this respect, with a symbolic equation being drawn increasingly strongly between what experts call 'antiscientific fundamentalism' and public ignorance of biotechnology, and embryo research and implantation. Indeed in this latter area experts have expressed the fear of persecution by 'mob rule'.[5]

Yet a significant feature of many contemporary risk issues is that polarisation only becomes more extreme as the experts engage in elaborate public information programmes to explain and reiterate their conviction that the risks are insignificant. Faced with this contradiction, the usual response is to intensify the programme rather than reconsider the assumptions on which it was based. Locked into this self-defeating cycle, many experts feel that society is being pushed irrationally into making decisions according to how many people will be frightened rather than how many will be killed.[6] Allied to the questionable belief that the media are both corrupt and all-powerful in shaping attitudes, this worldview is extremely resistant to falsification.

A field where the cycle of polarisation between experts and the public has been most pronounced has been that of nuclear waste disposal. Besieged by solid public refusal even to countenance exploration for shallow disposal sites, the then Secretary of State for the Environment, Kenneth Baker lamented in March 1986 that 'public anxiety over nuclear wastes has reached a level of hysteria to match that of the witch hunting crazes of the middle ages.[7] While experts in this field more than any other have been forced by experience to accept officially that public perceptions have to be taken into account, the conviction is still unchanged that they have no substance, and that whatever regular patterns they show to psychological research, they still have an inferior status to expert, real knowledge. Thus risk perception research, while it may have a useful role in explaining the curious beliefs of the public, is not thought to reveal anything that changes the moral standing of the respective beliefs, nor the political or institutional implications.

Risk perception research, as one of its pioneers Harry Otway has noted,[8] originated in the belief of its founders that the factors causing public 'misperception' of nuclear risks had to be identified and 'corrected'. Although many research findings have contradicted the assumptions shaping this powerful belief,[9] and despite a more accommodating policy stance towards public perceptions, the fundamentally patronising conviction still prevails that the experts know the 'real' risks while the public have only perceptions of them.[10]

One point of particular relevance here is the observation, made since at least the late 1970s,[11] that the public's assessment of risks is quite rationally based on judgements concerning the forms of control of those risks; this includes institutional judgements of the performance, attitudes, openness and overall 'social demeanour' of the relevant industries and regulatory bodies; in turn this logically requires empirical scrutiny of past institutional behaviour in these respects. These are not extra aspects distorting objective risk perceptions — they are judgements of factors that determine the objective risks. Equally, in attempting effectively to define these factors out of the risk analysis frame, experts are not avoiding institutional judgements in their technical analysis; consciously or not they are incorporating assumptions that these institutions will behave in a perfectly trustworthy manner. It is just such divergent but unarticulated institutional judgements and commitments which I argue inflate and confuse controversies about risks, and the responsibilities for the confusion seem to be more or less evenly distributed.

I will therefore challenge the predominant definition of the overall problem of expert–public disagreements about risks by (a) offering a reinterpretation of public risk perceptions that grounds their rationality in social experiences and processes, and (b) analysing the inevitably social framing of expert technical risk assessment. In the present chapter I focus on the second of these. The core of my argument is that the dominant perception of the overall problem — as essentially that of how to defend scientific rationality against disorderly subjectivism — is fundamentally flawed because of a common misconception about the nature of scientific rationality (and about public 'irrationality') embedded in our risk managing institutions. In particular it takes no account of the differences which can exist between real world risk systems and the models of such practices which scientists assume in order to structure and analyse a risk problem. It is precisely in this gap that ordinary lay social experience and judgement are pertinent to the analysis of risks. What is taken to be an irrational rejection of scientific definitions of risks, indeed of science *per se*, may be seen instead as rejection of naïve assumptions of an ideal world of operation, inspection, management or maintenance which scientists (following a more general mode of practice within science) have embedded in their risk analysis. As a general matter, as the following case studies show, institutions may make different social assumptions of this kind, and arrive at very different assessments of risks or regulatory decisions, even though they all appear to be following proper technical methods.

Below I describe four case studies which show the significance of social assumptions about the risk-system that are embedded in expert analysis. I then briefly consider the practical implications for decision-making bodies concerned with risk management, especially in the lengthening list of areas in which differences between expert and public risk perception exist. I argue more broadly that digging out the buried social assumptions in both expert and public discourse about risks and technology, and finding the means of debating them openly, is the central challenge if we are to progress beyond the current sterile predicament in technological risk management. The official recognition that public perceptions of risks count for something in risk decisions is a good start, but there is a further need for experts to be aware of and be analytical about the context-bound nature of their own rationality. Both experts and laymen need to be encouraged into processes which articulate their tacit assumptions, and allow negotiation between them; this means institutional as well as attitudinal development. The main work of establishing authentic dialogues between experts and laymen has yet to begin.

Case histories

Case one — 2,4,5-T

In the late 1960s the safety of various pesticides and herbicides was called into question. The herbicide 2,4,5-T, 'agent orange', was one of the foremost of these. In Britain the National Union of Agricultural and Allied Workers (NUAAW) lobbied for a ban of its use, bringing forth cases of medical damage to users in forestry and farm work. Through the 1970s the official government Pesticides' Advisory Committee (PAC) conducted several inquiries and issued eight reassurances that there was no scientific evidence of harm. The forestry and farmworkers' arguments were dismissed as unscientific and largely anecdotal. Stung by these rebuffs, and encouraged by a growing international movement of criticism, the National Union of Agricultural and Allied Workers collected more systematic case histories and evidence of unsafe practices on farms and in forestry use. Eventually in 1979 a further, fuller PAC review was conducted, but again a ban was rejected on the grounds that no scientific evidence of harm could be found.[12] This review however, contained crucial caveats: 'The use of brush killers containing pure 2,4,5-T offers no hazard to users, to the general public, to domestic animals, to wildlife or to the environment generally, *provided that the product is used as directed* . . . The use as directed of brush killers containing 2,4,5-T contaminated with TCDD (dioxin) poses no risk *provided that the maximum allowable level of the TCDD contaminant is not exceeded*' (italics added).[13]

The standard way of interpreting this conflict was either that the workers were being irrational, imagining harm where none existed, or that the experts were cooking the scientific books under discreet political pressure either to keep 2,4,5-T on the market, or to maintain their own prior credibility (or both). In fact all these claims were made by different sides. While I do not want to suggest that neither of these factors existed in some degree, they are not adequate explanations, and they are intrinsically provocative and polarising. The far more significant point was that after several scientific assertions of no harm, the PAC's further reassurance was eventually qualified with the apparently minor conditions, that 2,4,5-T was indeed safe, *so long as it was produced and used under the proper conditions.*

This was precisely the point at issue. The experts on the PAC were all toxicologists and related scientists, examining the scientific literature describing laboratory tests of toxic effects (usually short-term lethality) from controlled doses of the pure chemicals on specially bred animals. Their 'objective' risk analysis thus assumed a world outside the laboratory which was essentially an idealised social world in which farmers and farm-workers were always able and willing to use the chemical according to prescribed conditions of preparation and use: such as full instructions and the time to read and follow them; correct doses; the correct base solvent; full protective equipment; proper pressurised valves and spray nozzles; spraying only in certain climatic conditions. As a model of the real world, these assumptions lose all credibility. Even if the scientific judgements, within their own artificial, taken-for-granted context, were entirely justifiable sustaining ideal and uniform conditions in the real world is totally unrealistic. As one workers' representative[14] said, 'it was like asking someone to work in a laundry, but to keep out of the steam'. The PAC were described by one farm workers' researcher[15] as 'living in cloud-cuckoo land behind the laboratory bench'.

The point of general importance shown by this case is that different parties — the scientists and the workers — defined different actual risk systems, or risk-analytic problems, because they built upon different models of the social practices creating or controlling the risks. The scientists' implicit assumptions were of idealised worlds of herbicide production and use; and the validity and credibility of their 'objective' risk analysis was committed to this naïve sociology embedded in their technical analysis. Conversely the workers, whose risk perceptions were for a long time dismissed as overactive imaginings of side-effects, had real empirical experience, indeed expertise, that was directly relevant to an objective risk analysis.

The social assumptions or models shaped the expert risk analysis, but not at a level of deliberate choice or 'bias'. It was more a reflection of the structural role of scientists in the decision-making system, along with the narrowness of scientific education. However the inability or unwillingness of the experts to recognise the need to make social assumptions of the kind that were eventually exposed, and their automatic adoption of a patronising objective-perceived risk framework, only encouraged the view that they were out to defend the pesticides industry. Thus the expert advisory committee for risk management, and the institutions associated with it, lost public credibility (over a wider front than one issue) by reiterating the objective-perceived risk language for dealing with public concern.

Although it shows some interesting minor differences, the next sample indicates that the structural incoherence identified in the 2,4,5-T case may be of wider relevance.

Case two — animal growth hormones

In 1984 the European Commission asked one of its expert advisory committees to examine the safety of five hormones proposed for production by bioengineering methods, for increasing yields in the beef industry. The Lamming Committee reported in October 1985.[16] Its conclusions were that two fully synthetic hormones were of uncertain safety and should be subjected to further research. Three nature-identical growth hormones, however, were judged to be acceptable under certain conditions. These included specific dose limits, non-edible site of injection, full veterinary supervision and a minimum waiting period before sale of treated meat. The Committee decided that if these conditions were met, the residues of the injected hormones were not detectable above the background natural levels of beef cattle.

On this scientific basis the European Commission wanted to permit the use of the three nature-identicals and thereby also encourage the biotechnology industry. However, intense political lobbying against the idea was successful in forcing the Council of Ministers to reject their experts' advice and instead they banned the hormones. A Directive to that effect was issued in December 1985.[17] In the immediate aftermath of this political decision, indignant experts in the Commission expressed outrage at the way in which, as they saw it, the 'scientifically well prepared' proposals of the Commission had been swept aside by a wave of ill-informed, emotive popular superstition.[18] The most senior Commission directors were asked to tell their political masters what a disgraceful travesty of rational decision-making had been allowed to take place. The message was that industrial civilisation could not survive if this kind of irrationality was to be tolerated. Yet let us look more dispassionately at the structure of the issue.

The Lamming Committee declared that the hormones were safe if certain practical conditions of implementation were met. The Commission staff apparently did not stop to ask whether these conditions were socially realistic to enforce, on hundreds of thousands of farms, across the twelve member-States of the Community, before advocating commercial production and use of the hormones. It is reasonable to interpret the political opposition as a tacit judgement about the social enforceability of precisely those conditions which the Commission staff ignored, yet which were crucial in determining whether or not the hormones risks would be kept beneath a 'natural background' threshold. In this case at least, the scientific committee did mention the conditions which would have to prevail in the system for their risk assessment to be valid. However the policy experts took an idealised model of the social world for granted in constructing their 'objective' view of the risks. The political opponent, it can be suggested, took an equally objective but opposite view of the practical world of enforcement, and effectively judged the policy experts' assumptions to be incredible. Yet again there was no recognition of this social or institutional dimension to the risk assessment and decision. Instead there was only polarisation and self-righteous escalation of the experts' paranoia about so-called public irrationality. Where there could at least have been a relevant debate (no doubt not without conflict) there was irrelevant and degenerative posturing.

As with the 2,4,5-T case therefore the 'objective risk–subjective perception' version of the whole issue can be exposed as not only deeply flawed, but also socially provocative. The public is being asked to place faith in experts who tell them they are stupid, irrational or at least gullible, when the experts themselves may on occasion be pronouncing 'objective' risks based on optimistic fantasies about behaviour in the real world. Within given conditions, people do of course get risk-estimates wrong but research has shown that their ability to make quantitative estimations is not that bad;[19] it is certainly not bad enough to explain the wide differences of risk judgement evident in the kinds of case described above. Divergencies of *social* judgement of the kind I have suggested have to be introduced to explain this.

In the next case I show a similar structure in an issue which (apart from sporadic outbursts) has not as yet escalated to national levels of polarisation and conflict between 'experts' and 'the public', but which may already be well on its way there.

Case three — the landfill of hazardous wastes

The United Kingdom's official national policy in waste disposal espouses the practice of 'codisposal' of 'special waste' (e.g. toxic) with ordinary municipal wastes in controlled landfills.[20] This practice is widely dispersed in about 1,500 different sites, controlled by nearly 200 waste disposal authorities (county and district councils). This 'dilute and disperse' practice is controversial internationally because it is regarded as a less rigorous as well as cheaper method of industrial waste disposal. It allows undue competitive advantage as well as uncertain long-term environmental and health protection. Because of its low cost it is an attractive proposition for import of wastes even across the Atlantic and several such plans gained hostile attention in 1988. Other countries (such as the United States and West Germany) have chosen to try to reduce dependence on landfill, to ban more materials from landfill or to adopt concentration–containment methods when landfill is unavoidable.

The British system *is justified* on the grounds that as a system it is far from indiscriminate.[21] Indeed it is extremely elaborate involving controls at site-selection, site-licensing (which can include restrictions on certain waste types, volumes and physical forms), management codes of practice and inspection. These controls are performed by the waste disposal authorities who are assumed to be in the best position to optimise regulations (e.g. licence conditions) to local conditions (e.g. site characteristics, waste types and so forth).

Codisposal was scientifically justified as a general policy in the mid-1970s by research at several sites which showed that attenuation of leachate by natural clay lining usually offers protection of groundwater. However, this technical risk assessment depends critically upon the diverse practical realities of local implementation as outlined above. In *theory* it offers a far more refined, discriminating system of risk management than crude blanket bans or fixed national standards. Instead of these indiscriminate standardised approaches it can optimally match particular wastes to particular sites, their particular environments, specific forms of treatment or pre-treatment and local disposal practices. In *practice* however, the system has been found to be failing drastically to fulfil those conditions of implementation.

Whereas the official policy is defended by the Department of the Environment by assertions that 'sensible landfill of special wastes is acceptable' on scientific and public health grounds,[22] the crucial question is whether 'sensible' landfill is actually enforceable in the extremely dispersed, heterogeneous, free-market British waste treatment and disposal industry. The Hazardous Waste Inspectorate, set up in 1984 after the House of Lords (Gregson) Committee's criticism, itself issued successive Annual Reports outspokenly critical of local authority implementation of national regulations. A Royal Commission referred to the situation[23] as 'ramshackle and antediluvian'.

Site selection, licensing and inspection have all been found to be so bad that the optimal 'overall risk assessment' based on the assumption of an ideal local enforcement world is seen to be quite unrealistic. Whether an adequate approximation to the assumed ideal social world can be achieved in the near future by tightening up but avoiding rigid national standards, is open to doubt. In any case it is possible that what is in theory a very refined technical risk management framework will eventually be forced to take more account of the real social world on which its validity depends. In this case the national system may be forced to build regulations and risk management on the assumption of considerable 'underdevelopment' of analysis, inspection, monitoring, etc. at local levels thus imposing more restrictive blanket conditions on landfill to avoid reliance on universal good practice at so many diverse sites, with diverse operators and local authorities. This is effectively the situation in the United States where different social assumptions about local site practices have led to more standardised, precise and restrictive criteria regarding landfill, and more comprehensive and rigid design criteria for such sites.

Elements of the same polarisation, outlined earlier between experts assuming that their framework of assumptions produced an 'objective' risk picture while the public's is irrational, have already arisen in the hazardous waste field, but so far with less paralysing overall effect than in, for example, the nuclear waste case.[25] Already, however, policy analysts have warned that the evident inadequacy of practical implementation threatens to undermine general public confidence in hazardous waste regulations, institutions and their experts. The analytical point I am making is that a regulatory policy commitment to allow variation and discretion in interpreting standards into practice — to optimise the risk management of the system overall — depends critically on the assumption that local practice interpretation is really capable of fulfilling that responsibility everywhere.

I will now turn to a rather different type of case to illustrate that similar framing problems to a greater or lesser extent pervade the full field of technological and environmental risk management and the relations between expert and layman.

Case four — the Abbeystead explosion

When the underground valve house methane explosion occurred in the Lune-Wyre water transfer scheme in 1984, killing sixteen people and maiming twenty more, the Health and Safety Executive conducted an inquiry into the causes of the explosion.[26] Questions raised ran from the advisory (what lessons can we learn for dissemination to others so as to avoid repetition?) to the accusitory (is there evidence for bringing a charge of criminal negligence against anyone?). The most common question was, who if anyone, was to blame? Was a methane explosion not foreseeable, and therefore avoidable, given the state of expert knowledge?

The HSE found that neither the tunnel designers, constructors nor operators could have been reasonably expected to have foreseen the emission of methane in the tunnel, and that therefore no one was to blame. Even before they had reported however, various interested journals[27] were busy publishing evidence which *prima facie* appeared to contradict this judgement, including published descriptions stretching back to the 1930s, of lethal explosions in apparently similar systems. It also included accounts of widely available local knowledge about coal measures and old surface coal workings in the area close to the tunnel. Coal is naturally associated with methane. In its inquiry Report however, the HSE omitted to even mention much of this evidence of prior knowledge, let alone explain why it was not judged relevant. This silence appeared to suggest to some that the HSE had at best been incompetent, and at worst, may have been engaging in a cover-up to defend the tunnel engineering industry and its reputation.[28]

Our research on the Abbeystead issue has suggested an explanation which preserves the integrity of the HSE, though exposing a lack of institutional self-awareness. We suggest that the HSE had examined the issue from the standpoint of just one of its several statutory functions, namely prosecution for criminal negligence. In framing its inquiry, it had asked itself the question, should anyone involved be so prosecuted? And it had examined and sorted evidence from this stance. The point is that a criminal negligence charge (rightly) requires a severe burden of proof — 'beyond reasonable doubt'. It also effectively asks the question: did the responsible party (designer, constructor or operator) *know* that methane was likely, and then ignore it? This is a very different question from: given the state of knowledge existing in the engineering profession (and amongst lay people) at large, *should* they have known? In gathering evidence for the first, more severe, criminal test, the HSE ignored evidence which was possibly relevant to and strong enough to answer the question of civil negligence, or even to common-sense notions of what the experts might have been expected to know, or at least watch out for.

The key point is not that the HSE should have not asked the criminal negligence question, but that it does have other statutory functions, which involve gathering evidence and analysing lesser degrees of failure, in order to learn and disseminate better knowledge and practices for the future. Excluding this information from the report effectively undermined these functions, especially since nowhere in the report was it made clear that for particular and temporary purposes the higher (criminal negligence) threshold of knowledge was being used. Thus a particular institutional scenario, criminal prosecution, was used to structure an analysis which was presented as an objective analysis, not one based upon rather special and limited criteria for defining facts and prior knowledge.

In this case we can again see two opposing rationalities working themselves out into an increasing general corrosion of public credibility of a major government institution concerned with controlling risks. The HSE's view of its own credibility is that it will only prosecute firms if it is virtually certain to win the legal case — to bring a charge then lose it is to lose credibility, and credibility with the companies it regulates is essential in a UK regulatory culture which is mainly informal, confidential and consensus-orientated. Yet it is this very rationality which has created an impression of cover-up, bias or incompetence to another part of its spectrum of social relationships. Cover-up and conspiracy do not have to be invoked to explain the HSE's partial construction of the facts in the Abbeystead case. It can be put down to the unreflective working of an institutional habit which has been allowed to settle down and shape what then seems to the experts to be a natural, objective scientific framework. By this lack of institutional awareness, public trust (and indeed that of independent experts) in the HSE was jeopardised.[29]

Conclusion

In all cases of real-world risk management, whether retrospective inquiries or prospective regulatory analysis, experts have to make many assumptions about the social and technical nature of the system whose risks they are trying to analyse and control. When we analyse the environmental risks of placing an oil pipeline through Northern Canada for example, do we assume the proposed pipeline will or will not be the first element of a whole corridor of developments? And do we restrict our analysis of construction damage to that done under official norms of winter only, or do we suppose that under pressure of deadlines, for example, the active season will inevitably be stretched into the longer, warmer days of summer, when the natural environmental is far more fragile?

Even relatively precise mechanical system risk management has to assume a certain level of manufacturing competence, maintenance thoroughness and inspection.[30] Right down to this detailed level therefore, scientists have to make social judgements and commitments which define how a risk-system is to be analysed. Some of these may be broadly agreeable, but others may not. Also, they may affect 'macroscopic' system properties 'does this reprocessing plant 'require' another, or commercial plutonium use, to be economic? Can this herbicide's or growth hormone's use be adequately socially controlled on a large scale? or 'microscopic' properties 'are those data from a reliable source? Do they apply to the case in hand?'

Whatever the degree of acceptance of each inherent social assumption in the 'objective' frameworks, the important point is that the more such frameworks are advocated as the only rational knowledge of the risks, the more they invite critical scrutiny. And the more vigorously their objective pretensions are broadcast, the harder they and associated institutions fall when, inevitably, their foundation on social assumptions is eventually exposed by that critical scrutiny.

The assumptions which analysts make are often an unconsciously expressed function of their own social roles and relationships within the system. For example, toxicologists used to highly controlled laboratory conditions and using idealised concepts with which to develop theory, are not well fitted to appreciate the importance of real-world practices that diverge wildly from ideal models. Experts in decision-making institutions find it hard to appreciate how someone completely outside the process may feel about that institution's accountability and commitments.

Experts' assumptions may be so unrealistic that to make their 'objective' risk analysis valid would require reorganisation and control of the world in the image of the laboratory experiment or the computer model, with all their implicit and explicit framing restrictions and conditions. For example, if pesticide risk assessment assumes ideal conditions of use the response to more realistic accounts of practices may be to 'discipline' use to bring it more under control. In principle this is sensible (as has happened with pesticides in the United Kingdom) so long as the feasibility and other consequences of this extra social control are carefully weighed.[31] Some of these tacit expert assumptions are about concrete parts of the social world with which ordinary people either themselves have direct experience, or about which they are as much entitled to judge as anyone else. If the risk analysis and the risks are sensitive to such factors, as they often are, then to dismiss public concern incredulity or opposition as irrational is directly provocative and self-defeating.

Indeed even if on analysis the risks turn out not to be sensitive to such differences of social judgement or definition, the very fact that the expert framework ignores them even as questions is a source of legitimate public anxiety. At the moment we have little in the way of institutional mechanisms for identifying such conditional assumptions and evaluating their range of effect on the risks, their uncertainty or instability and what we should do about them. Instead such framework questions emerge, if at all, in the destructive context of political polarisation, bad faith and anxious laments from experts about being engulfed in a new dark age of anti-science, superstition and hysteria. I do not want to belittle the dangers in such a scenario. However, I suggest that it is the very attitude behind such laments which provokes the response we wish to avoid.

In overall conclusion therefore I emphasize that even the most controlled, objective knowledge is embedded within a tacit framework of idealised, fixed relationships. This is what allows it to *be* controlled knowledge. But in unreflectively extending from the laboratory or computer model to make their observations about the real world, scientists rest the validity and the objectivity of their statements upon the built-in assumptions about that real world and the social shape and limits of the real risk system. The 'objective' framework floats on a sea of subjective commitments and assumptions which have to be more openly expressed and negotiated in risk assessment processes. Lack of awareness and realistic assessment of these assumptions not only risks getting the 'objective' risks wrong; it also causes socially provocative assertions to be made about the moral worth of the relationships and realities of ordinary people. In their innocence therefore, scientists peddling the 'objective risk' framework as a decision-making standard are unwittingly sowing confusion and conflict, when they think they are the last defenders of enlightenment.

To escape this downward slide we do not need any more assertions of the 'objective risk–public hysteria' framework. What we need instead is to nurture frameworks for getting 'inside' the expert analysis, to identify its framing social assumptions and allow them to be examined critically in a measured and open-minded way. It is no coincidence that the technology which is suffering most from 'extreme' public reaction is the one which for its own legitimation has previously gone out of its way to 'black-box' itself in images of white magic and impenetrable expertise.[32] There has never been a means of more measured criticism short of indiscriminate opposition and unfocused fear. This is surely a far more important, expert-imposed social factor in the extreme public anxiety around nuclear power than the fashionable talk of the unseen nature of radiation. Has anyone ever seen lead or mercury in the environment?

Before their evident value is lost in a welter of public disaffection, the need is therefore for a change in outlook on the part of the experts, to recognise the fundamental bounding of their own objective frameworks, corresponding with such a changed outlook however, is a need for changes in organisation of experts in the decision processes. The analysis of this chapter leads to the general conclusion that unexpressed assumptions prior to structured technical analysis are the important source of conflict between experts and laymen. These assumptions — for example about the feasibility of standards of operation, use, inspection, etc. — have to be more regularly and more openly expressed and negotiated in the risk management decision process. In this way we could advance the quality of public debate rather than allow it to founder in confusion and bad faith.

It may be that through the current confusion we will eventually see that we are engaged in a process of renegotiation of the proper relations between experts and layman. If so, this will not happen in an institutional vacuum, disconnected from social and political arrangements. Social scientists must lend their expertise in helping to fathom this new terrain, and to ensure its constructive mapping.

Notes

1. Wynne, B., 1984, 'Public Perception of Risks', in Surrey, J. (ed.), *The Urban Transportation of Irradiated Fuel*, Macmillan, London; Wynne, B., 1985, 'From Public Perception of Risks to Cultural Theory of Technology', in Covello, V., *et al* (eds), *Environmental Impact Assessment, Technology Assessment and Risk Analysis; Contributions from the Psychological and Decision Sciences*, Springer Verlag, Berlin, New York, pp. 849–75; Wynne, B., 1983, 'Redefining the Problems of Risk and Public Acceptance: the Social Viability of Technology', *Futures*, **17**, 1–29; Thompson, M., 1982, 'Outline of a Cultural Theory of Risk', in Ley, E. and Kunreuther, H. (eds), *The Risk Analysis Controversy: and Institutional Approach*, Springer Verlag, Berlin; Douglas, M., 1985, *Risk Acceptance According to the Social Sciences*, Russell Sage Foundation, New York; Short, J., 1984, 'The Social Fabric at Risk: Towards the Social Transformation of Risk Analysis', *American Sociological Review*, **49**, 711–25; Rayner, S. and Canter, R., 1987, 'How Fair is Safe Enough?, The Cultural Approach to Societal Technology Choice', *Risk Analysis*, **7**, 3–9.
2. Royal Society of London, 1983, *Risk Assessment: a Study Group Report*, Royal Society, London; and 1985 *The Public Understanding of Science*, London, Royal Society, London.
3. Rogers, Karen, 1987, of Monsanto, in *Biotechnology: the Challenge*, Proceedings of the Department of Agricultural Biotechnology Challenge Forum, Washington, DC, February, Part VII, 'How Can Biotechnology Reach the Public?'.
4. For analysis of this use of rationality as symbolic politics, see Wynne, B., 1982, *Rationality and Ritual: the Windscale Inquiry and Nuclear Decisions in Britain*, British Society for the History of Science, Chalfont St. Giles.
5. Professor Robert Edwards, at a meeting on Biology and Ethics, Venice, May 1988. Quoted in *The Observer*, 15 May 1988, 'Scientists Warn of Mob Rule'.
6. Fremlin, J.H., 1984, *Power Production: What Are the Risks?*, Adam Hilger, Bristol and Boston.
7. Baker, Kenneth, 1986, *Hansard*, **94** (14) March.
8. Otway, H., 1985, Introduction, Otway, H. and Peltum, H. (eds), *Regulating Industrial Risks*, Butterworths, London.
9. For a review, see Otway, H. and Thomas, K., 1982, 'Reflections on Risk Perception and Policy', *Risk Analysis*, **2**, 69–82.
10. Thus the title of a 1987 paper to the Safewaste 1987 Conference, Churchill College, Cambridge, March–April from Strange, R.C., UK Nuclear Industry Radioactive Waste Management Executive (NIREX) official was 'Public Perception and Real Risk'.
11. Wynne, B., 1980, 'Technology, Risk and Participation — the Social Treatment of Uncertainty', in Conrad, J. (ed.), *Society, Technology and Risk*, Academic Press, London.

12. UK Pesticides Advisory Committee, 1979, *Review of the Safety for Use in the UK of Herbicide 2,4,5-T*, Ministry of Agriculture, Fisheries and Food, Pesticides Division, London, March. See also, 1980, *Further Review of the Safety for Use in the UK of Herbicide 2,4,5-7*, December.
13. Ibid. (1979), p. 5.
14. Anon, 1980, quoted in *The Landworker* (the journal of the National Union of Agricultural and Allied Workers), June, p. 7.
15. Clutterbuck, C., 1980, in *The Landworker*, June 1980, pp. 8–20.
16. *Rapports du comité scientifique vetérinaire, du comité scientifique de l'alimentation animale et du comité scientifique de l'alimentation humaine sur la base du rapport du groupe scientifique sur les agents anabolisants dans la production animale*. 1984, Commission of the European Communities, Brussels, Directorate General Agriculture, EUR 8913, (1984).
17. Council Directive of 31.12.85, *Prohibiting the Use in Livestock Farming of Certain Substances Having a Hormonal Action*, Brussels.
18. I am grateful to Mark Cantley, of the European Commission's Concertation Unit for Biotechnology in Europe (CUBE), for discussion and personal communication of these events and related issues. See Cantley, M., 1987, 'Democracy and Biotechnology, Popular Attitudes, Information, Trust and the Public Interest', *Swiss Biotech*, 5–15. The quote was noted from an internal commission memorandum dated January 1986.
19. Although lay publics tend to compress the margins by underestimating large fatality potentials and overestimating low ones, their ability to rank hazards according to quantitative harm is reasonable. Slovic, P., *et al*, 1979 'Rating the Risks: The Structure of Expert and Lay Perceptions', *Environment*, **21**, (April); and Slovic, P., 1987, 'Informing and Educating the Public About Risk', in Kunreuther, H. and Kleindorfer, P. (eds), *Managing and Insuring Hazardous Risks — from Seveso to Bhopal and Beyond*, Springer Verlag, London, New York, pp. 307–32.
20. For a fuller account, see Wynne, B., 1987, *Risk Management and Hazardous Wastes: implementation and the dialectics of credibility*, Springer Verlag, London, New York and Berlin, Chapter 7 (with E. Madel), 'Decentralised Regulation and Technical Discretion; the UK', pp. 195–243. Material for this section was gathered during research under an ESRC project, and I am grateful to Duncan Laurence, research fellow, for discussions on these issues.
21. UK Department of the Environment, 1981, written evidence to the House of Lords Select Committee on Science and Technology, chairman Lord Gregson, *Hazardous Waste Disposal*, volume III, London, HMSO.
22. Ibid.
23. UK Hazardous Waste Inspectorate, 1st Report (June 1985), *Hazardous Waste Management: an overview*, Department of the Environment, London; and 2nd Report (1986), *Hazardous Waste Management: Ramshackle and Antediluvian?*, Department of the Environment, London. See also Laurence, D., 1987, 'The Department of the Environment — Ramshackle or Antediluvian?' *Waste Management*, 77, 28–34.
24. Royal Commission on Environmental Pollution, 1985, 11th Report, *Managing Waste: the Duty of Care*, HMSO, London.
25. Paradoxically, we have argued, the same local control which allows wide variation of standards encourages a degree of public tolerance not afforded to the centrally controlled nuclear waste enterprise, even though risk acceptability–design criteria are usually more rigorous in the nuclear waste case. Centralising and standardising non-nuclear waste regulation would not necessarily lead to

greater public acceptance — indeed it could do the opposite. Wynne, B. and Laurence, D., 1988, 'Physical Risks, Institutional Structure and Public Reactions — a Comparative Analaysis of Nuclear and Non-Nuclear Waste Management in the UK', University of Lancaster ESRC-funded project, Draft paper.

26. UK Health and Safety Executive, 1984, *The Abbeystead Explosion*, HSE, London. This section is based upon collaboration with Lyn Aitken, Bob Lee and Jos Kent. I am grateful to Mr John Rimington, Director General of the HSE, for discussing the issue with one of our group (LA).
27. The *New Civil Engineer* published regular bulletins and reviews from summer 1984 to May 1985 reporting *inter alia* earlier experience of methane explosions in similar situations. What was 'similar' was a key issue in determining whether prior knowledge existed.
28. This idea was put to us by several interested parties with whom we talked during our research, and in some press accounts at the time.
29. Later developments in the Abbeystead case, after this paper was written, do not affect the analysis offered here. A civil case for negligence was tried in the Crown Court at Lancaster between January and March 1987. Mr Justice Rose presiding found the defendants — the designers, tunnel drivers, and operators — negligent on the grounds that there was inadequate coordination and communication between them concerning methane hazards and precautions. He allocated damages in proportions of 50 per cent, 35 per cent and 15 per cent respectively. The Court of Appeal in April 1988 rejected the designers' appeal, and increased their proportionate liability to 100 per cent, exonerating the others. Leave to appeal to the House of Lords was refused. In the meantime, the victims and their families had still received no compensation because this will have to be determined at future hearings into individual cases.
30. See for example, Ostberg, G., 1986, 'On the Trustworthiness of Safety Assessments for Nuclear Pressure Vessels', *International Journal of Pressure Vessels and Piping*, 25, 257–66.
31. Of course a regular problem is that technological commitments are made without prior attempts to analyse implementability and its safety (and other) implications. Then when divergences are found that make the original commitment questionable, it is too late to disentangle. Instead, the social world, for example, of work may be increasingly controlled and policed to try to make the real world comply with the original naïve social model(s). See Wynne, B., 1988, 'Unruly Technology — Practical Rules, Unpractical Discourses and Public Understanding', *Social Studies of Science*, **18**, 47–65.
32. Wynne, B., 1982, *op. cit.*, note 4.

Chapter 4

Risk Perception and the Public: Insights from Research around Sellafield

Sally Macgill

Introduction

The disputed connection between radioactive discharges from the nuclear installation at Sellafield in Cumbria, United Kingdom, and an apparent excess in the incidence of leukaemia and other cancers among children who live nearby, has generated one of the most significant public controversies there has been in recent years about issues which have at their heart perceived threats to health. In this chapter recent research into aspects of the Sellafield controversy is used to elucidate some of the real complexities encountered in coming to terms with the different perceptions of risk which lie at the core of this controversy, and others of a comparable kind, and which stand in the way of acceptable resolution. A systematic review of related literature is not attempted (reviews of selected examples can be found in, for example Douglas, 1986; Renn, 1985; Council for Science and Society, 1976; Royal Society, 1983; O'Riordan, 1983), rather, specific illustration of four key issues.

The chapter is written in contrariety to a technocratic[1] view of the problem presented by people's adverse perceptions of 'the risks' from nuclear technology — a view of ignorant and misinformed minds which are prey to deception and fear by irresponsible and unrepresentive political activists and the media. This technocratic perspective may be convenient and temporarily satisfactory for a 'rational' and 'knowledgeable' industry and regulatory authority (but see the companion chapter by Wynne (Chapter 3)). However it is plainly inconsistent with serious understanding of risk perceptions: it is inaccurate and misses issues of real significance. Moreover, its ways of addressing the 'adverse perception' problem have tended to be limited to increasing authoritarianism ('we know best'), doing nothing (because people are 'irrational' and 'immature'), undermining people's credibility (rather than addressing their concerns) or advocating programmes of education of staggering ambition ('if only people were better informed . . .'). Each in its way is unsatisfactory.

The basic premise of the present chapter is that what are ordinarily referred to as 'the public's risk perceptions' are of far greater moment than the technocratic view recognises. Rather than being dismissed, excused or so badly misunderstood, the complexion and determinants of people's perceptions of risk need to be explored more seriously. This must entail sound methodology and empirical evidence (not bogus techniques or pure speculation), in order fully to engage with the true complexities of people's perceptions of risk. Questions increasingly faced by policy advisers and decision-makers, such as that of how they might weigh scientific assessments of risk against public perceptions, can then be addressed from a more informed position.[2]

This chapter seeks to foster a fuller understanding of public risk perception; of some of the complexities encountered in related research; and of difficulties facing attempts to ameliorate related risk concerns and conflicts. This will be approached by sketching out some answers to the following four questions: (1) Who are 'the public'?; (2) What determines their risk perceptions and attitudes; (3) How can their risk perceptions and attitudes be represented?; (4) What might be done to ameliorate adverse and 'inflated' risk attitudes? The questions are addressed in the context of research into the Sellafield risk controversy (for a fuller account, see Macgill, 1987). It is hoped that the answers can provide useful new starting points for those faced with addressing and negotiating problems in which adverse risk attitudes play a significant role.

None of the four questions holds a simple or straightforward answer. The complexities of the first lie in the many different constituencies of people with a claim to interest in major risk controversies — lay, workplace and professional communities; private and public interests; active environmental lobbies and intentionally passive publics; powerful influences and politically unprivileged classes; the press and mass media — and by the absence of 'laws of social entitlement' *vis-à-vis* the weight that should be given to the views of different constituencies. The second involves consideration of the multitude of differentiated economic, cultural and social forces that shape the different ways people 'see' related issues, and the different positions they in turn adopt. The third is problematised by difficulties in faithfully representating people's views and positions, due to distortion and incompleteness in any representation. The fourth is confounded by difficulty in overcoming formidable social and historical forces that give rise to prevailing differences among different people's positions.

The context in which the above questions are addressed in the present chapter — disputed health risks from Sellafield — has been one of acute public controversy in recent years. Of particular significance was a television documentary broadcast in November 1983 ('Windscale: the nuclear laundry', Yorkshire Television) which drew attention to an unusually high incidence of leukaemia in children in the vicinity of Sellafield — ten times the national average in the under-fifteen-year-olds in Seascale, the nearest village. Radiation was noted to be the only known cause of leukaemia within the limits of present knowledge and Sellafield was observed to have discharged greater quantities of radioactivity than any other installation in the United Kingdom. The programme was heralded by a considerable amount of advanced publicity, due to a row between the management of the company who operate the reprocessing facility at Sellafield, British Nuclear Fuels Limited (BNFL) and the programme makers (Yorkshire Television). It was watched by an estimated three million viewers.

The speed of government response to the programme bears witness to the sensitivity and controversiality of issues forced into wide public view. Within 24 hours of its broadcast a Committee of Inquiry under the chairmanship of an eminent medical scientist, Sir Douglas Black, had been commissioned to investigate matters that had been so dramatically highlighted. The findings of the inquiry were published the following summer (Black, 1984), confirming an excess in the incidence of leukaemia among young people in Seascale, stating that a link with radiation from Sellafield was 'unproven', offering a 'qualified reassurance' to anyone concerned about a health hazard in the vicinity of Sellafield and making a number of recommendations for further research.

Who are 'the public'?

The first question is important because debate about 'public' risk perceptions can become frustratingly ambiguous, sterile or (perhaps deliberately) misconceived if the answer is not apparent. In commentary about Sellafield for example (and ambiguity over who really are being referred to as 'the public' is hardly exceptional to this case), different references to an otherwise unspecified entity simply alluded to as 'the public' turned out to be references to one of a number of different populations:

(1) The national audience of the YTV programme, people concerned about its disclosures and/or allegations.
(2) 'The people' of: West Cumbria, Seascale, Copeland, six parish councils, Millom ward . . . other spatially definable local areas.
(3) 'The people' of: Britain, Scotland, Northern Ireland, Eire, the Isle of Man.
(4) Regional Councillors, District Councillors, Parish Councillors, MPs . . . in other words, elected representatives.
(5) People who write letters to newspapers, are interviewed on TV in a non-professional capacity, who talk to journalists.
(6) Activists: Greenpeace, CORE (Cumbrians Opposed to a Radioactive Environment), FOS (Friends of Sellafield), Coastwatch . . .
(7) People who support particular political/interest groups (to the exclusion of people who do not).
(8) All but one or more of the following: politicians, nuclear industry management; nuclear industry employees; civil servants; experts; . . . other professionals.
(9) The media — very frequently taken as some sort of barometer of a broader, unobserved climate of public mood or opinion.
(10) National Mori/Gallop poll participants (70 per cent of whom,

according to a recent poll, considered Sellafield to be unsafe; 40 per cent of whom considered it should close).

(11) No one in particular.

Serious appreciation of 'public' risk perception must acknowledge that there are many different publics — a fact plainly neglected or left ambiguous in dominant technocratic views, as well as in opposition 'populist views', which instead falsely adopt an implicit absolute of a homogeneous body of people who need not be identified, still less consulted.

An example of how easy it is (inadvertently) to ignore considerations of which public is being referred to is found in statements made by Sir Douglas Black following the publication of his Inquiry's report. Sir Douglas repeatedly voiced his belief that 'members of the public' who had been concerned about possible radiation-induced health risks arising from Sellafield's operations would be justifiably reassured. It was not always clear which particular public he, or others who adopted the same position, had in mind. But empirical evidence — interviews with several hundred (N = 462) inhabitants of West Cumbria, reported fully in Macgill, 1987 — challenges this belief in respect of many of Sellafield's local public, as these examples illustrate: belief for some members of one particular important 'public' in this context (see Macgill 1987).

'. . . we've had many reports. Nothing has been done. There are still arguments. The powers that be should know what they're doing . . .'

'Don't think he [Sir Douglas Black] really settled it one way or the other — it still left you guessing . . .'

'Still left a doubt. We hoped there would be a conclusive answer.'

Though differences in attitudes, interest and value may be fully expected in different subpopulations, it should also be recognised that within any given subpopulation, it is possible to have considerable differentiation. In other words, simply identifying 'the public', is not to fix on a uniform mass of fossilised opinion. In Seascale, for example, a village virtually entirely dependent in one way or another on Sellafield, all inhabitants similarly stigmatised by recent publicity and all living in the village allegedly most directly 'at risk', there is not a simple homogeneous pattern of opinion. On the contrary, there are striking differences. There are about as many people who speak of concern for radiation-induced health risks linked to Sellafield's operations as there are who reject any basis for concern (see Table 4.1). The village is not in the grip of endemic fear; neither is it one in which there is little concern, expressed only by an insignificant, irresponsible or clueless minority, or where it is unanimously repressed by an employment conscious local workforce. Rather, there is a breadth and diversity in local 'risk attitudes', underpinned by rationalisations, justifications and arguments of differing nature, depth and complexity. There seems no compelling reason to suggest that diversity in attitude is peculiar to this particular subpopulation.

To get beyond the ambiguity about who really are 'the public' can raise

Table 4.1a How people describe their feelings about radiation induced health risks in the context of radiation levels of BNFL's operations at Sellafield

	Not worried	*Concerned*	*Anxious*	*Cannot say*
Personal risk	67%	22%	4%	7%
Risk to other adults	64%	21%	5%	7%
Risk to children	45%	40%	12%	4%

Table 4.1b How people felt about radiation levels for each of the following

	Not worried	*Concerned*	*Anxious*	*Cannot say*
On the beaches	45%	39%	12%	5%
In house dust	75%	9%	3%	13%
In the air	69%	16%	4%	12%
In fish and other sea food	60%	26%	8%	7%
In milk	71%	14%	5%	10%

Table 4.1c Concern about radiation induced health risks to children, according to people's place of work

	Concerned or anxious	
Place of work		
Public administration	72%	more concerned
Home	60%	↑
Sellafield	45%	↓
Retired	29%	less concerned

considerable difficulties which it may be politically convenient not to lay bare. Such difficulties include the realisation that one public cannot necessarily be assumed to be representative of another, and perhaps the onus to discover what the various publics' positions really are, and if they are in line with prior assumptions. Acknowledgement of the manifest diversity of people's positions can also prompt the uncomfortable realisation of the impossibility of pleasing everyone. So who matters most? Is it those 'objectively' deemed to be at most risk? Or those 'objectively' deemed to have most at stake? Or those with the most adverse perceptions? Or those whose interests are in most conflict with those wanting to understand perceptions? Or those most politically dangerous? Or those fitting some more general notion of a 'climate' of opinion? It would be naïve to pretend that there are ultimate guidelines for determining such issues.

Given the practical difficulty of keeping track of all possible publics, choice becomes inevitable. The way this choice is made can, in the end, only be determined by identifying ones motive for wanting to understand public perceptions. Possible motives are many and they include: keeping on top of the politics of anxiety; seeking to change (re-align) people's perceptons; ameliorating concerns; or making a disinterested study of a phenomenon to elucidate its intricacies or possibly to test related hypotheses. While disinterested academic research may identify most closely with the last of these motives, agencies more centrally and actively involved within risk controversies (having, in some sense, a 'stake' in any outcome) are more likely to be motivated by one of the earlier mentioned goals. What is clear is that the choice about 'which public?' can be problematic (perhaps a subject deserving of research in its own right), and its outcome will inevitably direct attention to a particular public and exclude others. Research findings, deductions or conclusions about public risk perceptions which may then follow may have very little relevance for other publics (for example, special interest groups of wider communities; elected representatives of smaller-than-constituency subpopulations) — a fact that is often ignored.

The intricacies over who really are the public are unimportant if technocratic pretences can be maintained. They can continue to be ignored by, for example, media commentary, which can continue to sustain itself on less rigorous an appreciation of reality. Greater integrity is required on the part of anyone seeking a serious understanding of perceptions. In summary, 'the public' consists of a multitude of overlapping and diverse subpopulations within and between which there are very different attitudes. Clarification over motives for wanting to comprehend public risk perceptions should direct attention to particular subpopulations, and any specific findings that may follow may have little relevance beyond these groups.

What determines people's risk perceptions and attitudes?

At a general level, this question can be answered by reference to Figure 4.1. It portrays risk perceptions as being determined on the one hand by individuals' own private reflective understanding of things — accumulated and assimilated over a lifetime of personal experiences — and on the other by individuals' wider continuing communication and interaction as a member of a social milieu. This includes both informal (for example, friendship, kinship) and formal (for example authoritarian, workplace, commercial) networks, whereby communication and dialogue with others can serve to confirm and verify positions already held or perhaps lead to revisions in view. The model presents people's perceptions, attitudes and opinions (or, more generally, their positions) as products both of their own awareness and interpretations of things and of their material and social (cultural and institutional) experiences as members of a society. Suggested components of these experiences for people in West Cumbria are sketched in Figure 4.2.

Apart from interpersonal communication and interaction, which will be

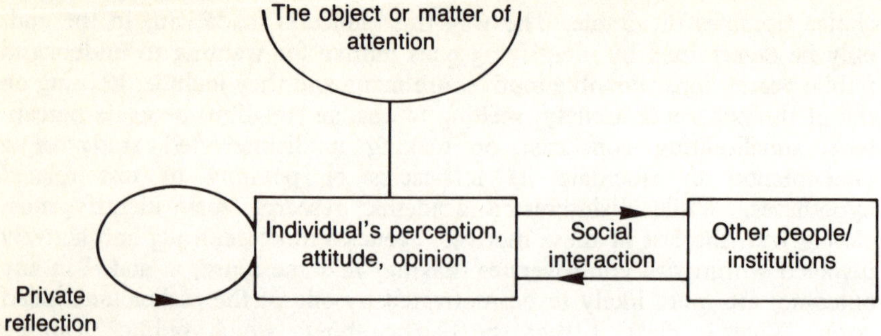

Figure 4.1 Model of risk perception process

elaborated more fully below, the dominant determinants of risk attitudes in West Cumbria (subpopulation 2 in the previous list) would appear to be the following:

(i) Material benefit — the importance of Sellafield to individual and community prosperity is very much recognised.
(ii) Personal investment — it is no less locally apparent (though much less appreciated more widely) that individuals and communities have invested their whole lives in Sellafield, and this cannot but condition the way they look at the issues.
(iii) Workplace — with a Sellafield worker in virtually every family, everyday experience of operations/containment/malpractices/pride or dissatisfaction with work is very readily apparent, and group knowledges quickly spread and disseminate to others and become embedded in the local vernacular.
(iv) Malign effect — many (though not all) locals recognise their area as one which has high cancer/leukaemia rates — a phobia of the twentieth century.
(v) Science — different understandings of science and different views of the authority of science and science-based control of the nuclear industry.
(vi) Media — either providing continual proof of Sellafield's malign radiation environment; or doing increasing harm to local quality of life through their stigmatising allegations.

The different mix of these factors generates the differentiated local views that have already been referred to; and while some people's attitudes rest on very firm and self-confident justifications (steadfastly rejecting any opposing forces), others signal people's confusion and disorientation.

Broadening the discussion to a wider British public (subpopulations 3 or 10 in the previous list) the various influences on people's perception of nuclear risk have clearly produced adverse effects. A plausible explanation for this might run as follows. Media commentary, for many people the sole means of awareness, feeding on operational errors and scientific disputes

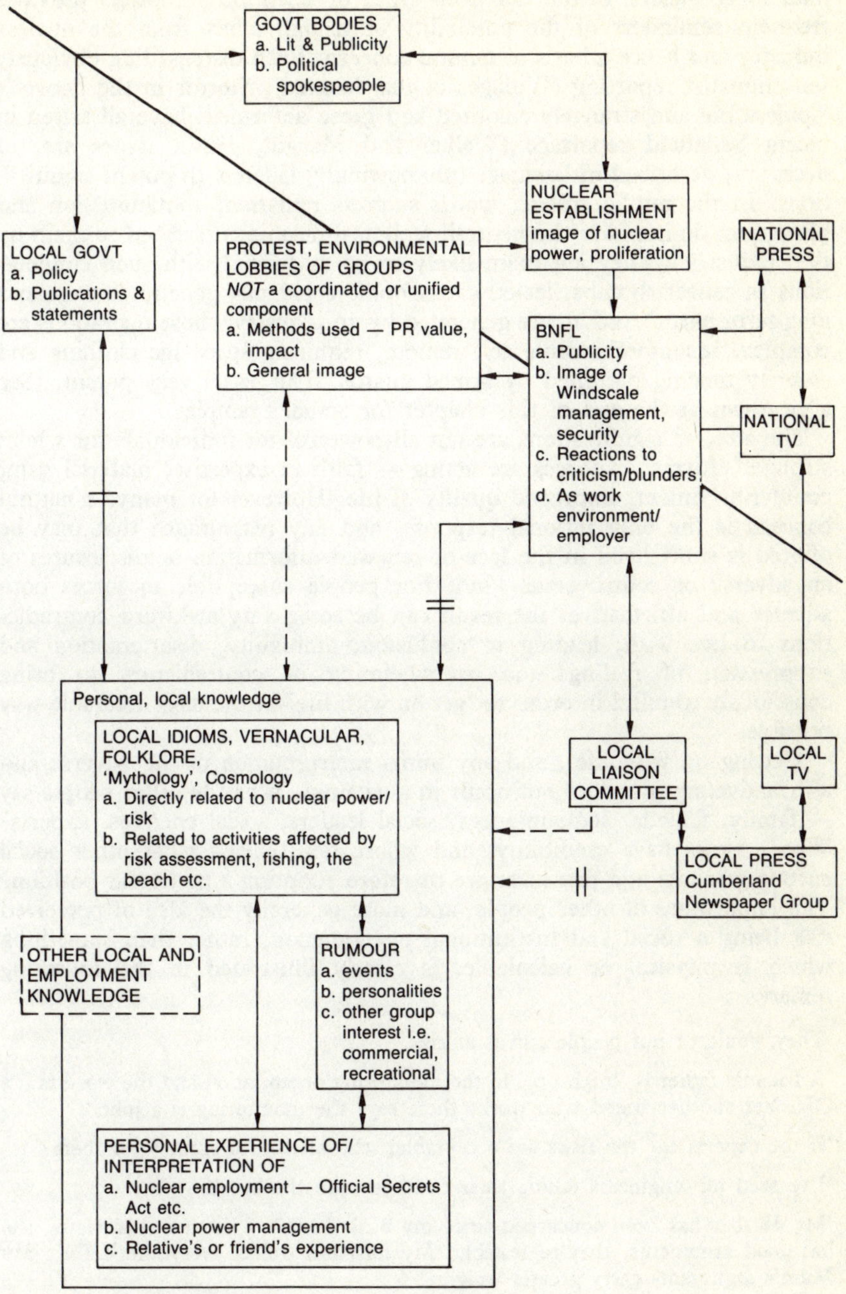

Figure 4.2 Networks involved in risk perception

(and more generally, the fall from grace of scientific authority) provides frequent reminders of the possibility of malign effect from the nuclear industry and hence a basis of natural concern. Even disregarding obviously sensationalist reporting ('Villages of the damned', 'horror in the hoover', 'spiders big and strangely coloured and geese deformed' have all arisen in recent Sellafield reportage (Walker and Macgill, 1987), issues are, of necessity, described in language (unknowingly) laden with potent significations. In the public sphere, words such as radiation, contamination and plutonium do not have the neutral, technical meanings ascribed to them by the 'industry'. They are more likely to be associated with such connotations as cancer, bombs, leaks, catastrophic accidents, genetic damage and long-term harm. And all are generated by an industry whose operations are complex, incomprehensible and remote, requiring space-age clothing and security fencing patrolled by armed guards. This is all very potent. (See appendices at the end of this chapter for actual examples).

The adverse significations are not all-powerful for individuals for whom stronger affirmative forces are acting — faith in expertise, material gain, counter-argument, improved quality of life. However for many, a natural concern is the only rational response, and any reassurance that may be offered is short lived in the face of renewed information or disclosures of an adverse or controversial kind. For people susceptible to forces both adverse and affirmative, the result can be some very awkward contradictions to live with, leading to confusion, ambiguity, disorientation and suppression of feelings too overwhelming or contradictory to bring consciously to mind in order to 'get on with life' in the most tolerable way possible.

'Getting on with life', and any initial interpretation of the adverse and affirmative signals, does not occur in a vacuum. What do other people say — family, friends, acquaintances, social leaders, social enemies, experts? Whose views have credibility, and whose don't? And what other social encouragements and pressures are therefore adopting a particular position? The importance of other people, and more generally the idea of perceived risk being a social and institutional phenomenon, rather than something which is physical or calculable, is clearly illustrated in the following remarks:

'They wouldn't put people's lives at risk.'

'A friend's father is "high up" in the monitoring organisation and the word is it's OK. Yet another friend who works there says the monitoring is a joke.'

'If the experts say the risks are acceptable, who are we to argue with them?'

'I've seen top engineers taking their children onto the beach.'

'My Mother has been concerned since my birth in '58. I listen to her views, she has good arguments, they're feasible. My husband thinks everything's OK. My Mum's arguments carry greater weight.'

Social interaction leads to positions being confirmed, adapted and, in some cases, discovered or expressed for the first time. Positions and arguments which are 'successful' get repeated and so both an individual's own rationalities and the cohesion of particular social groups is maintained and, likewise, the rejection or suspicious appraisal of opposing or threatening positions. Prejudices develop. These are not irrational whims that may be ignored by anyone with a serious interest in understanding perceptions or that can be suppressed by an individual. Rather, they are forces which act as filters to people's openness to the world and are a part of their very being. They are simply conditions whereby what people experience or encounter says something to them. Prejudices — filters of perception — can be very deep rooted indeed and have very deep social investments within them: systematic mistrust of authority; systematic faith in employer or in science; the automatic ascribing of particular meanings (and not others) to words.

It is not simply a question of risk perceptions and attitudes deriving from interpretations which are shaped by people's already held prejudices and which are reinforced in the course of their daily lives as social individuals. Rather, risk positions themselves can become a means for developing social cohesion and interrelationships — the feeling of belonging to a community or cause group, perhaps (see also Edelstein, 1986). It also then becomes obvious and rational behaviour to take on board whatever other arguments and possible activities reinforce the group's aims (and corresponding extreme derision or hostility towards threatening ones). The 'risk perceptions' that are the forces of internal cohesion for some groups can at the same time be forces of dissent and disruption in some broader social context.

What is also important to appreciate is that people's own prejudices are not always fully apparent to them — and this includes 'experts' and 'authority' as well as 'public' (see again Wynne (Chapter 3)). Thompson (1980) has talked of people being ensnared in their own provincial perspectives. Linkages with the ways in which others in 'the same' controversy look at things may be tenuous or unknown. To ask 'who is right?', for example, about how big the risks really are, may be a proper focus for inquiries concerned only with the validation or refutation of objectively verifiable 'facts'. But that is not the problem posed by adverse risk perceptions. It does not help to tell people they are 'wrong'. Analysts who fail to recognise this, but speak only of 'irrational' bias and 'misperception' miss issues of real interest and mask them with their false implicit absolute of rationality. Rationality in the present context is perhaps most usefully defined with respect to the prejudices which constitute it. There are many.

The perception analyst, though having to confront the manifest complexity and diversity of people's prejudices and ways of looking at risk issues, is at least relased from the onerous, and now evidently impossible, task of searching out an unbiased, unprejudiced or uniquely rational view. In summary, people's risk perceptions are determined by their interpretations of what they recognise as the attributes, material and symbolic, of the risk source (in the present context, the nuclear reprocessing facility at Sellafield). What attributes people identify and what they then mean depend on people's prejudices and values and their experiences of everyday, economic and social, life. They are manifestly different for different people, and fluid. To speak in terms of irrationality and misperception, among other elements of self-blinding vocabulary, is plainly to misunderstand how people, all of us, cope with conflicting realities.

How can risk attitudes be represented?

Whereas in answering the previous question (what determines people's risk perceptions and attitudes?), the various influences on people's perception of risk have been surmised, here it is necessary to define what representation will be used as evidence of what people's perceptions are. In particular, what sorts of data, measures or other characterisations can be used to tell an analyst, or other interested observer, that there is or is not an 'adverse risk perception' problem. There are many possibilities. Any one of the following might be referred to as an indicator of social attitudes to Sellafield:

attitude polls;
qualitative scalings;
representation of local discourse;
products of anthropological observation;
political action (for example, Greenpeace);
evidence of local protest action;
the size of subscriptions to particular campaigns;
news media reportage;
media correspondence;
media documentaries;
fish sales;
hotel use;
beach use;
house prices;
parliamentary questions;
letters to MPs;
children's drawings;
waiting lists for particular GPs practices;
a sense of crisis.

Against this array of diverse possibilities three points in particular will be chosen for elaboration.

First, each individual indicator will give its own characteristic depiction of risk attitudes. None can claim to do so in a comprehensive or permanent way. In other words, none should claim to be truly 'representative', though some studies may claim otherwise, but only partially and temporarily. No one would claim a painting to be a comprehensive representation of its subject, a mathematical model a comprehensive representation of the system it depicts or a map a comprehensive representation of the earth's topography. The same is true of attitude representation. A representation of the social world cannot (by definition) in itself be the social world. One can seek to observe and somehow record particular social practices —

people's behaviour, arguments or whatever — or identify some likely indirect indicators (fish sales, hotel use). Some will yield more faithfully, telling and longer-lasting clues than others. What is represented will depend on what is chosen to be observed and on the instrument of observation, including how it is employed (for example, the design of survey questions, choice of sampling frame and sensitivity and guiding intuition of the analyst). It is also important to recognise that particular indicators are, either by nature or by design, specific to particular publics. It is no practical use to intend to 'monitor' the perceptions of a particular public using an indicator which reflects another. For example, West Cumbrian attitudes cannot be represented through national opinion polls; nor can moral support for political activists be measured solely via a count of active members or of subscriptions.

Second, there is the question of which representation to refer to. It follows from the previous point that there is no unique or objectively correct choice. While it is possible to visualise an ideal choice process, in which particular (and most faithful) representations are consciously chosen from amongst all possible alternatives, in practice this ideal is unlikely to be realised. The most faithful representations may not be the easiest to obtain — notably because of inevitable resource constraints — or the most obvious (for example the politically most active groups may assert their importance over others). There may be political pressure on the analyst (or other commentator) to adopt a particular representation in preference to others. Or a particular representation might be unconsciously selected through the investigator's own disciplinary biases or subcultural environments and adopted as if it is the only possible reality. It is not at all clear that perception researchers have adequately managed the problem of representation, but its existence should at least be recognised, and it should not be dismissed as being insignificant.

In summary, although representations are inevitably partial and selective, faithful representations of risk perceptions must be grounded in people's own experiences and interpretations. Anything else, or the blind adoption of a single representation, to the exclusion of others, is make believe, therefore of limited validity and may even be counterproductive.

What might be done to ameliorate adverse risk perceptions?

Adverse nuclear risk concerns delineate a crisis of confidence in the industry, with ramifications for those with statutory responsibility for their regulation and management on behalf of society. To overcome the crisis would require forces to overcome those which are currently acting to shape attitudes in an adverse way. Quite simply, if 'industry' and 'attitudes' are out of kilter, their (re)alignment requires either (a) changing attitudes, or (b) changing industry (or a bit of each). What are the possible means?

Changing attitudes

(a) Two strategies for changing attitudes are: (i) to induce people to change their positions by offering them material gain; (ii) by reason and argument. Each can be problematic.

(i) The elementary logic behind the first is plain: if people stood to gain more obvious benefit, they may take a more positive attitude. It is the logic behind suggestions to offer compensation to communities finding themselves local hosts of noxious dumps (see Puschak and Burton for an early proposal); it is also behind oversimplistic and somewhat inaccurate risk-benefit explanations of the apparent acceptability of risk to Sellafield employees.

The logic is too elementary for general application, however, there being ready evidence that the attempted substitution of money for value can be pathetically inadequate and sometimes quite counterproductive. First, the logic that for some individuals constitutes a reasonable risk-benefit balance is for others plain evidence of harm — '. . . the compensation paid out by BNFL to the families makes me feel that there is something in the accusations against them' — or immorality: '. . . it's disgusting to try to buy off people's lives like this'.

Second, the complex differentiations within society mean that the 'average' individual for whom compensation may be 'calculated' is unlikely to be representative of many others, and may not even exist. Third, the material compensation strategy tries (falsely) to re-focus the perception problem back onto risk as a quantifiable phenomenon, or as a set of finite, discrete attributes. Since people do not see the problem in such terms but rather in a much more complex and often ambiguous way (cf the answer to question 2, p. 49), so the logic for risk-benefit compensation can become very tenuous indeed. Fourth, even without the above complications, the practical achievement of compensations that are generally seen to be fair could be quite problematic.

(ii) Reason and argument is not necessarily more successful. The message of reassurance, emerging from Sir Douglas Black's high-level Committee of Inquiry, and intended to assuage people's concern about possible environmental health risks from Sellafield, was received in West Cumbria with a mixture of disinterest, confusion, satisfaction, deferred judgement and sharp criticism. Overall, the reassurance and the argument on which it was based, though repeatedly stressed by Sir Douglas in a large number of mass media statements following the publication of his Inquiry's report, did not appreciably affect the way most people in West Cumbria look at the issues. Notwithstanding the point that 'presentational' aspects could have been better,[3] and the possible obscuring effects of the scientific controversy that followed the Black Report's publication,[4] this is uncomfortable news for anyone who would want and expect such a distinguished public policy initiative (and argument of reassurance) to have greater social relevance. Its role otherwise is merely to interest scientific, medical and other elite communities, to exist as an object of occasional media attention or, more hollowly, to stand simply as a ritual act of political expediency (Macgill, Ravetz and Funtowicz, 1985).

That Sir Douglas was a medical man, and a somewhat anti-establishment figure with no connections with the nuclear industry, really underscores a profound problem here. In particular, the House of Commons Environment Committee (1986), among others, have talked of the importance of having experts who are independent of the nuclear industry to offer service and reassurance to members of the public. Here was an important occasion when such independence of expertise was nowhere nearly enough.

The above West Cumbrian example is by no means unique in pointing to the lack of effect of authoritarian reason and argument. Drawing on a wealth of academic experience (and perhaps also their own) the Environment Committee (1986), for example, plainly recognised the enormity of the problem, in commenting that it is naïve to think that the public as a whole is easily going to be 'educated' technically in the way the industry has imagined, nor are their adverse perceptions likely to be countered by superficial public relations' tactics. Once more it is important to acknowledge the formidable complexity, subtlety and ambiguity of people's perceptions.

And, of course, there can be no promise of the 'reason and argument' strategy succeeding among people who are convinced about immaturities and uncertainties in the science of radiation-induced health risks. For these populations, there is no scientific basis for offering reassurance. It is also only sensible to acknowledge that the distinction between 'public' and 'authority' among participants in related scientific debates is not necessarily that between 'lay' and 'expert'.

The limited effectiveness of reason and argument can perhaps be more closely considered by discriminating between two (linked) aspects — the social and the scientific. In social terms, limitations arise due to stronger forces acting at interpretational, psychological, cultural, institutional, experiential or political levels (these are the sorts of things described in answer to question 2, p. 49). In scientific terms, limitations arise if there are legitimate points of entry (because of immaturities, uncertainties and the fundamentally socially constructed nature of knowledge (Ravetz, 1971)) for fundamental dispute of the scientific basis of reassurance: in other words, dispute over 'the facts' about radiation-induced health effects from nuclear power.

Succeeding in changing attitudes through reason and argument is difficult enough if the possibility of legitimate scientific limitations can be discounted. The task is then that of trying to find a final compromise between people's conflicting interpretations, wants, beliefs and values, but at least against an assured basis of scientific fact ('truth' about how big the risks really are). The difficulty of the task is severely compounded if the possibility of legitimate scientific dispute cannot be discounted — a position for which there is a growing weight of opinion and evidence, not least Chernobyl (also, perhaps, the prospect of increasing debate as a result of increasing access to information; and new research findings such as those potentially resulting from the studies instigated by the Black Report's recommendations). There are then not only the contradictions and fluidity of the social world to contend with, but also the complications and ambiguities of an uncertain (though not non-existent) science too.

So, whereas in the presence of social, but not scientific, limitations there is the possibility, however 'irrational' to certain minds, that social choices will reject what is scientifically sound; in the presence of possible scientific limitations, the possibility of adopting a course that proves folly both in scientific and in social terms becomes all too real. The question of what should be done in such circumstances is not easy to answer, though to act in pretence that the science is definitive when it is actually embryonic is to fudge the issues, to invite later rebuke and to renounce the principle of knowledge-based decision-making (see Funtowicz and Ravetz, 1987, and also their chapter (7) in this volume, for new ways of representing uncertain scientific knowledge).

Changing industry

(b) Alternatively, rather than seeking directly to change public attitudes, possibilities for changing the industry, thereby making it more publicly acceptable, might be pursued. These could act on two levels: (i) change the industry's operation; (ii) change the industry's style.

(i) The proportion of the present crisis of confidence is in so small way due to the number of operational incidents (obviously Chernobyl and Three Mile Island, and in the United Kingdom, not least Sellafield). The beliefs of an increasing number of people, that errors are systematic or 'normal', mean that for them a future of stable operations (one possible view of 'change') is just wishful thinking. They instead advocate more radical change — running the industry down. A parallel state of affairs is in evidence in the context of low- and intermediate-waste disposal — whole communities calling for a radical change from the proposed option of shallow trench sites.

For some the entertainment of such radical alternatives flies in the face of science;[5] for others their dismissal flies in the face of reason. What, if anything, should be done in terms of change, then, depends on how much authority should be given to (which) science, reassurance and its institutional practice; and how much to alternative social wants. The choice will depend, not for the first time, on motives, as will the thoroughness with which the costs, opportunity costs and error costs of different alternatives (in all their complexity and subtlety) are investigated.

(ii) Changes of style, as interpreted and recommended by the Environment Committee (among others), and gaining increasing Political currency, would mean the industry ridding itself of its remoteness, over self-confidence, complicatedness, secrecy and unaccountability. Such changes would constitute something of a revolution in terms of the kinds of social contract under which the industry has operated throughout its life to date, and it is difficult to see the extent to which the nature of the nuclear industry is intrinsically compatible with such change. After all, the crisis has deepened at a time when the industry has moved closer to 'the public' than ever before.

In summary, to seek to change public attitudes is a formidable prospect. A failure and counterproductivity of past initiatives can be explained in terms of their failure to catch the depth, complexity and ambiguity of people's perceptions. The way ahead must be to learn to live more positively with, on the one hand, the realities of the social world and the demands it makes on, in this case, the nuclear industry and regulatory authorities and, on the other, various uncertainties in the state of scientific knowledge. In a democratic society, this entails not imposing industrial or authoritarian over social wants, and not disguising or dismissing imperfections in knowledge. Rather, in a period of uncertain transition, it is to be more receptive to calls for new social contracts and institutional reforms whereby the newly emerging demands and newly appreciated uncertainties may be more realistically confronted.

Conclusion

A crisis has been described[6] as a moment of transition in which something which held before does not hold any longer, and there is an inadequate capacity of society to respond to change in a positive way. This description seems very fitting to the contemporary position of the nuclear industry in the United Kingdom. Tacit acceptance on the part of a large number of publics has been successively replaced by growing doubt and recognition of uncertainty; growing concern; and accidents, incidents and fresh 'disclosures' about the health hazards of nuclear power. This is at the same time a portrayal of change, and of far from positive adaptation. The crisis is manifest not merely in terms of such symptoms as low opinion poll ratings, and increasing receptivity to 'bad nuclear news' reportage, making for a lack of acceptance of nuclear power, and difficulty in communication between 'industry' and 'public'. More seriously, there is outright rejection of the industry by an increasing number of publics, total collapses of communication and, seemingly, less and less agreement as to what any future dialogue might be about, or who it should be between.

Any valid weighing of scientific risk assessment against public perceptions in nuclear risk management will have to catch the currents of the contemporary crisis. Anything else would just be a misplaced desire for (a socially rejected) utopia.

The answers to the four questions posed in this chapter have tried to contribute to the ways in which the 'adverse nuclear risk perception problem' — and crisis — may be sensibly contemplated. They amount to a statement that it has a depth and objective complexity which must be more honestly acknowledged and reckoned with. If the appearance is that of so much complexity and so few final answers, then some further comments are perhaps in order. One is that serious social research into nuclear risk concerns is not the soft, unimportant option that it is often made out to be. Rather it is a point of entry into worlds of conflicting social interpretations and wants and uncertain scientific knowledge. Another comment, possibly more profound, is that in more completely discovering problems — revealing complexities that have not yet been sufficiently appreciated and thereby fostering wider acknowledgement of how difficult certain problems might be to confront realistically — the academic can claim to be engaged in a singularly appropriate and positive activity. To have more effectively forewarned policy advisers about the comprehensive rejection by communities up and down the country of sites contemplated for nuclear waste might be an example of where a superficially negative 'academic' appraisal is in reality just what was needed to be known. Only on the basis of appreciating the true nature of current or impending problems and crises can developments which seek amelioration of them be properly founded.

Notes

1. 'Technocratic' denoting faith in the performance and 'rightness' of, in this case, nuclear technology.
2. This premise was also the very germ of the meeting for which the material of this chapter was originally prepared: a seminar in London in July 1986 involving officials from the Department of the Environment and scholars engaged in risk perception research.
3. People commented that Black was distant; that the report was difficult and that summaries were not available.
4. For example, a succession of articles in *Nature*, the *Lancet*, *New Scientist* and the *Guardian*.
5. There were comments to this effect in opposition to proposals for zero (or in some cases just reduced) discharges from Sellafield.
6. Discussion between U. Eco and S. Hall, 'Voices' Channel 4 Television, May 1985.

References

Black, D., 1984, 'Investigation of Possible Increased Incidence of Cancer in West Cumbria', report of an independent advisory group, HMSO, London.

Council for Science and Society, 1976, *Of Acceptable Risk*, Barry Rose Publishers Ltd, London.

Douglas, M., 1986, 'Risk Acceptability According to the Social Sciences'. Russel Sage Basic Books, New York.

Edelstein, M., 1986, Phychological impacts of toxic exposure; an overview in Becker, H.A., and Porter, A.L. (eds), *Impact Assessment Today*, Volume 2, Jan Van Arkel, Utrecht, 761–776.

Funtowicz, S.O. and Ravetz, J.R. (1987) 'The Arithmetic of Scientific Uncertainty', *Physics Bulletin*, **38**, 412–14.

House of Commons Environment Committee, 1986, *First Report, Radioactive Waste*, House of Commons Paper 253, i–xvii, HMSO, London.

Macgill, S.M., 1987, *The Politics of Anxiety: Sellafield's Cancer-Link Controversy*, Pion, London.

Macgill, S.M. and Berkhout, F.G., 1986, 'Understanding Risk Perceptions: Conceptual Foundations for Survey-based Research', School of Geography, University of Leeds.

Macgill, S.M., Ravetz, J.R. and Funtowicz, S.O., 1985, 'Scientific Reassurance as Public Policy: the Logic of the Black Report', Working Paper 448, School of Geography, University of Leeds; also in Macgill, S.M., 1987, op. cit.

O'Riordan, T., 1983, 'The Cognitive and Political Dimensions of Risk Analysis', *Journal of Environmental Psychology*, **3**, 345–54.

Ravetz, J.R., 1971, *Scientific Knowledge and its Social Problems*, Clarendon Press, Oxford.

Renn, O., 1985, 'Risk Perception: A Systematic Review of Concepts and Research Results', paper presented to the meeting on 'Risk and Policy Analysis', International Institute for Applied Systems Analysis, Laxenburg, Austria, November.

Royal Society, 1983, *The Assessment of Risk: A Study Group Report*, Royal Society, London.

Thompson, M., 1980, 'Political Culture, an Introduction' WP-80-175, International Institute for Applied Systems Analysis, Laxenburg, Austria.

Walker, G.P. and Macgill, S.M., 1986, 'The Black Inquiry and the Media', working paper, School of Geography, University of Leeds; for revised and reduced version see Macgill, S.M., 1987, op. cit.

Wynne, B., 1986, this volume, Chapter 3.

Appendices

Appendix 1 Languages of risk in West Cumbria

'I'm bothered when I go past the plant — I feel I ought not to breathe'

'The workers at Sellafield are probably the healthiest in the area'

'I'm terrified; every time I have an ache, the place comes to mind'

'It's wrong that so much was shoved out to sea, you hear a lot from workers there that a load more gets shoved out at night — when nobody's looking. It happens all over, not just here. They seem to let go twice as much as they should do'

'Can't really blame BNFL as my husband works there'

'I'm at the sharp end so I know what I'm talking about, radiation hasn't done me any harm in the four to five years I've worked in the radioactive area'

'They talk of working within limits — but it's a new industry — who makes the regulations? It's so new they don't know who it's going to harm'

'I'm quite anxious most of the time. It makes you more concerned when the children are ill'

'I lived here when I was a child and I'm alright'

'You always seem to have a thought in the back of your mind if you do go on the beach, you wonder if you should be on it, not perhaps for yourself, but your children'

'If I was concerned, I wouldn't have retired here. Why aren't there dead fish and birds floating about?'

'I'm a bit worried about future generations'

'Cars kill more people than radiation'

'I live each day as it comes — you've got to die of something'

'You can never say you're not worried about radiation problems . . . and someone who is not worried, in a sense, is daft'

'I'd still go on the beaches . . . wouldn't swim though. There may be a lot of doubt there . . .'

'It's a cloud hanging over us — so much uncertainty'

'I'm surprised how many women are worried but don't admit it because their husband works there'

Appendix 2 Some illustrative risk positions, by social group

Retireds
'If there was a danger from cancer, I think I would have contracted it by now'

'I might be more worried if I was younger or had children'

[Also, an age of greater deference to professional expertise and the perfectability of nuclear technology]

Sellafield employees
'I'm at the sharp end so I know what I'm talking about. Radiation hasn't done me any harm in the 4–5 years I've worked in the active area'

'My knowledge of the plant makes me confident that BNFL is working satisfactorily. It's just the November 1983 issue that brought it into people's minds. They don't realise you can get a higher dose of radiation just from luminous watches'

'Jake Kelly shouts about as if he's running Sellafield. He knows nothing about the place. Got to work there to know'

[Mutually reinforcing exclusions and justifications of working at Sellafield. 'Concern' as a word not much in evidence in vocabulary]

Home
'There's nowhere for children to go in summer. You could turn a blind eye but you're never sure. My husband wouldn't let the children on the beach'

'You always have a thought in the back of your mind if you do go on the beach, you wonder if you should be on it, perhaps not for yourself but your children'

'I can't really blame BNFL as my husband works there'

'Different experts have given me totally opposing views . . .'

[All housewives; resp. for children; only indirect knowledge of Sellafield; rely on views of others; statement of unconcern; defensive of husband's workplace]

Public services etc.
'I feel I shouldn't be subject to *any* levels of radiation and I resent someone saying 'it's safe' because I feel no-one really knows'

'They talk of working within limits — but it's a new industry — who makes the regulations? It's so new they don't know who it's going to harm'

Images of disaster
(top) Sinking of the Titanic (bottom) Chernobyl nuclear reactor fire

Dark ages? No thanks
Atoms for energy!

CEGB
SIZEWELL POWER STATION
PRIVATE ROAD
NO ADMITTANCE TO PUBLIC
NUCLEAR EXHIBITION &
INFORMATION CENTRE
OPEN TO THE GENERAL PUBLIC
ADMISSION FREE
MONDAY - THURSDAY 09.45 - 11.45 14.15 - 16.45
FRIDAY 09.45 - 11.45 14.15 - 15.45
SATURDAY - CLOSED
SUNDAY CLOSED
CAR PARKING NEAR CENTRE
FOLLOW THE SIGNS

Representation of the nuclear case

Public protest at proposals to site low-level radioactive waste

Chapter 5

Biotechnology: Anticipatory Risk Management

Jerry Ravetz and Jennifer Brown[1]

Introduction

This chapter identifies some of the hazards associated with the development of the biotechnologies, especially the release of genetically engineered micro-organisms. Previous case histories presented in this volume have dealt with contemporary or retrospective controversies. In contrast the study of the threats posed by biotechnology is very much anticipatory as risk assessment in this area is still in an embryonic state.

Collingridge (1980, 19) notes that it is desirable for policymakers to have accurate forecasts of the effects of a technology before it is highly developed and diffused. If done effectively, untoward consequences can be avoided and desirable outcomes enhanced. However, he not only believes that techniques of forecasting are dubious but also the prediction of social effects 'demands a vastly greater appreciation of the interplay between society and technology than is presently possessed'. The present authors are more optimistic. It will be argued that lessons can be learned from the experience of other technologies and related scientific disciplines and applied in an anticipatory way.

The first section outlines a programme for the development of a scientific and regulatory framework in which to safeguard health and harmony of people and their environment. The second section draws attention to the lack of public awareness of the risks and benefits of biotechnology and suggests some information dissemination and educational strategies to remedy this.

Managing the hazards of genetic engineering

Introduction

The study of the hazards of physical technologies, notably nuclear power

[1] The first section of this chapter was written by Jerry Ravetz and the second section by Jennifer Brown.

and chemical plant, is a substantial and well-developed field. Although like any field of policy-related research it is affected by many uncertainties and marked by public debate, it possesses a base in broad accepted data and theories and a body of skilled practitioners with their own organisations. In contrast, the study of the hazards of biotechnology is as yet in an embryonic state. The work began with a false start in the 1970s with the confused debates over laboratory-scale genetic manipulation. Scientists, particularly in the United States, came out of that experience bruised and disillusioned about public involvement in the regulation of science. Under those conditions, there was not much real progress in development of the ancillary science of recombinant-DNA hazards, necessary for the effective regulation of the research.

Now, ten years on, the atmosphere is matured and positive. In every quarter there is a determination that technical innovation and diffusion should be monitored, so that this new technology should be safe and be seen to be safe. There is a real concern for the fostering of research into the new hazards, and for striking the right balance in regulation.

Relevant past experience

The experience of other fields of study of biological hazards has varying degrees of relevance for biotechnology. The risks of medical and surgical interventions involve known procedures and a strictly limited population; similarly toxicology involves possibly harmful agents that generally do their damage only to persons actually coming in contact with an identifiable sample. Radiobiology is closer to our concerns, since (in part) it studies agents that diffuse into the environment; but these are of a few well-known types, which are relatively easy to monitor and whose biological effects (in all but very small doses) are well measured. Epidemiology and public health are closer still, and indeed form part of our concern, for they deal with self-reproducing agents that are not easily controlled once they are established in the environment. Also, the phenomena in these fields are not always simple, and can usually be studied only by the less rigorous methods of *post hoc* statistical analysis.

The fields which are more similar to the case of biotechnology have several properties that distinguish them from the established experimental sciences. The most obvious is the severe limits on quantification in their analysis. In the case of physical hazards and those similar to them, it is possible to construct models of the process whereby a hazard materialises, and these can then be interpreted mathematically. The parameters of the mathematical model represents measures of a transition from one stage along to the next, either a fraction of the input that survives the change, or a probability of a sample element doing so, or a probability of a discrete event that is necessary for further events downstream. The algebraic form of the simpler mathematical models is multiplicative (with the possibility of additive parallel pathways). Hence if several of the quantities in the series have a product that is in practical terms negligibly small, this component swamps the others; and the conclusion from the model is that the risk is in some sense small or 'acceptable'.

In the case of diffusing biological agents, and in particular microbiological ones, this convenient and comforting property is not available. It is well known in the case of virulent pathogens that a very small set of micro-organisms can constitute an effective inoculum, and once escaped can infect one and then many people. With agents of less severe pathogenicity, the situation is not so clear-cut; but again, a vigorous strain, once released into the natural environment, can have a subsequent history that is hard to predict and even harder to influence. In this respect, the microbiological hazards are reminiscent of the most intractible of the physical hazards, where probabilities seem vanishingly small but where harm, if it occurs, can be great beyond any assigned limit. Those hazards have been called 'zero-infinity', two terms whose product is indeterminate. Here the situation is even more complex, since the probabilities, in any realistic model, are strictly incalculable.

One useful way of characterising this difference is to think of biological agents as embodying information, rather than being simply harmful as individuals or collections. Information, as is well known, requires only a minimal component of mass or energy for its encoding or reproduction; and biological agents can propagate this information in a variety of ways, only some of them amenable to human intervention and control. This variety, and the associated flexibility, is useful to keep in mind when studying these microbiological hazards. We have seen how plasmids, viruses and even pieces of DNA, can migrate and transform independently of the larger entities which function as their carriers; and hence the pathways to the spread of information which is damaging are numerous and hard to define in advance. For this reason the assessment of diffused microbiological hazards must necessarily be something of an art as well as a science, depending on broad knowledge of the sorts of things that might happen, on the basis of examples that are appreciated as being of relevance to the particular case in hand.

Implications for practice

These characteristics of diffused microbiological hazards have certain consequences for the methods and style of the sciences by which they are controlled. These are similar to epidemiology, to an important degree proceeding *post hoc* to unravel the causes of phenomena we wish to control or prevent. To establish this point by a contrast, suppose that we wish to know whether a particular oncogene can be expressed in some situation of interest. We then go through a fairly straightforward procedure of defining the problem, specifying methods, obtaining support and permission and then testing for the effect under controlled laboratory conditions. At the end of the single experiment there are a few well-defined alternative conclusions, of course always relative to the state of the art of our techniques and theories, but otherwise capable of being presented for review and criticism. That is the science with which we are all familiar, and which has produced the great body of public, reliable knowledge on which we depend.

Coming to the case of diffused hazards, suppose that we want to know whether manipulation of a particular oncogene presents a hazard to laboratory workers. For many obvious reasons, there is no question of doing an experiment on people; hence we must have a survey. Now, that is a very different sort of undertaking altogether. In the absence of an effective regulatory body like the Advisory Committee on Genetic Manipulation (ACGM), obtaining such information would be far from straightforward. There are not merely the practical difficulties of finding out who has ever been in contact with the oncogene in question (including technicians, students and perhaps even cleaners and maintenance staff), and then obtaining their subsequent medical history. In addition, there is the complicating factor that such information is heavily value-loaded; managements will think twice before exposing their institution to a possible scandal through the actions of persons over whom they have no influence. Hence this simple question, which might appear analogous to the experimental problem, has a totally different character as a project.

Lest this example appear fanciful or invented for the purposes of an abstract argument, let us consider the case of the six cancers at the Institut Pasteur. If in 1985 someone had asked about the disease-rate in genetic manipulation laboratories worldwide, she or he would certainly have been told that there is no evidence of any additional hazard. And in a sense that would have been true, in that there was no public recognition of the four cases that had already occurred there. Indeed, it was not until late in 1987, when the cases had increased to six, and there were serious discussions within the Institut, that the scientific public had any published information on the problem. So if we were now to ask, how great are the hazards worldwide, the standard reassurance would not, or better should not, carry very much weight at all. But to obtain a reliable answer to the question, which indeed may come onto the agenda some time soon, would involve the mounting of a very large-scale international investigation.

This example illustrates the strong difference between the experimental disciplines that, in the course of their normal work, create the hazards problems, and the field/epidemiological disciplines that attempt to understand and control the hazards once they are created.

One important cause of the confusion and acrimony that afflicted the discussion of recombinant DNA research in the 1970s was this gap in ways of comprehending the problem. For the very successful experimentalists, scientific problems are, however difficult and challenging in their study, essentially straightforward in their structure. For those with a background in the sciences of environment and regulation, it is quite otherwise. The standard question in a debate, 'How do you know that . . .?' is, for experimentalists, settled by a reference to a largely public literature; for the others, it raises many problems of a methodological and practical order before an answer can even be framed.

Varieties of perceptions of uncertainties

This difference in perception of problems of knowing is associated with attitudes to the unknown that, quite naturally, tend to optimism among the laboratory scientists and to caution among others. In practice, in the American debate in the 1970s, this difference expressed itself as a policy presupposition: where lies the burden of proof on hazards, with the proponents of research, or with the opponents? Since the uncertainties in the situation were so great, this issue of methodology became the deciding factor in the regulation of the research. Of course the research was to be conducted with high standards of safety, within the framework of existing knowledge; but research on hazards would be minor, delayed and in effect concentrated on positive measures as the engineering of chi-1776. The opponents clearly recognised the imposition of the burden of proof on themselves and interpreted this as a self-conscious, self-serving political decision by the proponents. (From the other side, that was precisely the assessment made by the proponents on them.) However, it is far more likely that the proponents of the research were treating the hazards as just another challenging problem to be solved like all the others in which they had mastered the necessary techniques for success.

The confidence of the experimentalists in their ability to assess the risks of their work does not seem to have been misplaced; though the case of the Institut Pasteur may retrospectively put that in doubt, particularly if it turns out not to have been unique. And the ecological hazards of the possible release of special strains of micro-organisms are hard to see as anything but small. However, it is quite well recognised that large-scale operations are different, if nothing else than that the strains used must be robust or else they would not be viable on an industrial scale. The planned release of micro-organisms is now universally accepted as presenting problems requiring care and caution. Of course, there is experience to provide both reassurance and guidance. Particularly in large-scale processes, we have been using yeasts and bacteria for a very long time for our food and drink. And there are a few cases of systematic release, such as *rhizobium* and *B. Thuringiensis*, which have had no untoward effects. However, this experience is not the same as understanding, and there is much that we do not know in principle, and even more that we do not know in detail, about the ecology of micro-organisms in particular contexts.

Regulation under uncertainty

Perhaps the one thing of which we can be certain, is that there will be surprises for us when micro-organisms are released into the environment. No experiment or greenhouse trial can reproduce the diversity and variability of a natural context; and the behaviour of micro-organisms can depend critically on the sorts of conditions in its micro-habitat that are beyond the ability of science to predict or control. All we have at the moment, for general guidance, are some maxims of the sorts of things that are unlikely to occur. Thus, an organism which has been modified for some special function will in general be less adapted to survive in the wild. Whether it could transmute into a hardy, wild form may be doubtful, for that requires the cooperation of a number of genes. In general, the risks of release of modified organisms do seem to be reduced by such considerations, but these maxims admit of exceptions, and some case of ecological significance may turn out to be one such. For example, significant phenotypic changes can be triggered by the action of a single genetic event, as in the case of lepidoptera. Also, the distinction between domesticated and weedy or feral varieties of an organism is far from absolute.

The history of cultivars can involve repeated interactions and switches between wild, weedy and cultivated state. Even the humble carrot can become a weed, as in New England. As for invasive species, they usually fail, but sometimes succeed spectacularly for themselves and disastrously for the host environment.

It would be a mistake to think that the relevant uncertainties are restricted to the admittedly complex and ill-understood ecological side of the problem. Any time that we are assured that a particular procedure with DNA or viruses is safe in view of the known properties of these materials, we should remember how the 'central dogma' in whose terms the whole field was created has been falsified on so many counts, not least the presence and activity of retroviruses. Any reassurance based on a scientifically proved impossibility should be scrutinised for its 'pedigree': how frequently and recently has the consensus of experts changed on this topic?

Faced with such uncertainties, the regulators of planned-release experiments and industrial-scale applications can find themselves in a cruel dilemma. In some simpler cases, the intended use is of organisms that are not intended to survive in the wild, and in application these can be dead or really badly crippled (although on this there is never 100 per cent certainty). If the organism is intended to survive in its habitat, it must be sufficiently robust to meet the competition of the existing related biota. On the other hand, if the release is to be safe, the organism must not be so robust that it will displace existing species and/or migrate beyond its intended locale. To achieve this balance of strength and weakness would require fine-tuning of its properties, and prediction of its behaviour, to quite a high degree of precision. It is doubtful whether our knowledge is at present adequate to achieve such precision in many cases. Yet the regulator will always be under pressure to permit some such applications; if the burden of proof of safety on the proposer is too severe, then nothing will be permitted and the whole effort might as well stop before it begins.

Some might indeed argue that that is the only prudent policy, but such an attitude is both wrong in principle and unfeasible in practice. There will be experiments and releases, if not here then in some country with a less severe regulatory regime. And if every new technology were required to prove its freedom from harmful effects, then mankind would be very far from our present state. Of course, there is never a real choice between such absolutes; and the recognised task in the case of biotechnology is to strike a reasonable balance, always improving in the light of experience, between technical progress and environmental safety.

Lest this seem a somewhat anodyne recommendation, we must never lose sight of the stakes in this endeavour. At the moment, the technology and its problems are both embryonic, but within a few decades we can expect genetically manipulated organisms of all sorts being ready for release into the environment, accidentally or deliberately. This can possibly transform agricultural practice for the better, enabling a more 'natural' and less invasive and unstable means of supplying humanity with food, fibre and many other products. But when conducted on a large scale and with such variety of products and contexts, it presents the real possibility of ecological disruptions on a scale which make our present depredations seem benign.

In the light of that eventual promise and peril, we should consider how to design our approach to regulation so that it is appropriate to our present modest problems and also provides a proper foundation for the inevitably expanded effort of the years to come. At the moment, the pressure for experiments is still low, but our experience is quite scanty. Under these conditions there seem to be three components of an effective approach to the assessment and regulation of the hazards: ecological wisdom, public confidence, and the management of uncertainties. I shall deal with them in turn.

The relevance of 'ecological wisdom'

The prediction of the possible consequences of a release of genetically engineered organisms into the environment is very far from being an exact science. We have a very narrow base in controlled experience of such occurrences, and the general maxims that indicate buffers and barriers are by no means precise or universal. From an outsider's point of view, the situation might appear quite desperate: how can one possibly advise on the relative costs and benefits of an operation when its consequences not merely escape quantification but cannot even be classified with any confidence? It would seem that in such cases the prior attitudes would be dominant: with scientific optimism, one permits the experiment unless it is shown to be dangerous; with ecological prudence, one prohibits it unless it can be shown to be safe. Since the costs and benefits of each approach are in part hypothetical, and in any event likely to be long delayed in appearing, there is little possibility of testing and comparing the merits of the two approaches, in particular cases or in general.

Thus in this crucial area, the regulation of a scientific and technological process would seem to be outside the bounds of scientific knowledge and scientific method. It would be an extreme case of what Alvin Weinberg called 'trans-science' when he first encountered the problems of assessing the hazards of low-level ionizing radiation from civil nuclear power. This type of problem is indeed a major challenge to the science of our age, since it cannot but increase as the powers of our technology to transform and also disrupt the natural (and social) environment are steadily augmented. The attitudes and indeed the formal philosophies of science based on our heritage of success in research in the controlled conditions of laboratory research do not offer sufficient guidance in this case. They need supplementing by something new, or perhaps a new understanding of something old and now not so fashionable.

The reference to 'ecological wisdom' recalls an observation of Michael Polanyi about the sort of 'connoisseurship' that defined the knowledge of the old natural-history disciplines that flourished and indeed dominated science in the days of the gentlemen-amateurs. This involved the mastery, through years of experience, of a broad range of facts, so that every new experience could be comprehended in terms of precedents and analogues. In recent times, 'natural history' has acquired a pejorative sense, as if it were merely a heap of memorised facts. If we recall the aphorism about the history of human affairs, that it is 'philosophy taught by example', then we can appreciate that the history of nature can be no less philosophical or profound than the study of nature through the artificial conditions of the laboratory or through theoretical or mathematical analysis. The undeniable classic case of the union of history and philosophy of nature is that of Charles Darwin.

In the absence of a matured corpus of experimental results about the behaviour of particular organisms when released into the environment, an 'ecological wisdom' based on natural-history experience is the essential scientific approach to assessment and then regulation. On the basis of such an experience, there can be a reasonable and effective interrogation of a proposal. That broad background will provide assurance that no significant possible consequences will be missed and it will also filter out merely conjectural or speculative possibilities. The uncertainties can be listed, and classified and subjected to dialogue and eventual revision.

On the basis of such an 'ecological wisdom', there could be an effective monitoring and assessment of experiments involving releases into the environment; and so that aspect of the problem can be considered as soluable in principle. But there is a practical difficulty that must be recognised. This is that the concentration on the laboratory approach, to the derogation and eventual neglect of natural history in biological science, has had the effect of depriving us of the very skills we need in order to achieve effective control of the fruits of laboratory science and the technology it produces. By the end of this century, perhaps, we could face the paradoxical and also dangerous situation that we have a radically insufficient corps of scientists with the scientific competence to assess the hazards of the growing volume of environmentally intrusive genetic technologies.

There is no possibility of re-creating a class of gentleman-amateur scientists; and so the forecasting of an appropriate natural history will need to be accomplished by other means.

So long as a scientific career has a structure of employment and rewards that militate against the necessarily long and leisurely process of mastery of materials in a natural-history style, there will be few recruits to that approach. However should the problem be recognised at all levels of the scientific community, then there could be such a development, not along the old lines of course, but focused on environmental hazards and their comprehension and control.

The achievement of public confidence

A planned development of 'ecological wisdom' would interact with a second aspect: public confidence. One important cause of the troubles of the nuclear power industry in the United States was the sense among its critics, scientific as well as lay, that its scientists were afflicted by hubris. The risk problems that they had chosen to set aside as negligible were, in their view, the province of ignorant and irresponsible critics. A core meltdown was beyond the range of the 'credible' and so did not need to be catered for in their assessments. When it occurred at Three Mile Island, there was a drastic change in credibilities, with the meltdown gaining in status, and the scientists losing.

It is now more than twenty years since a public awareness of environmental problems became an issue for the conduct of industry and technological innovation. The confusions and extremism associated with the early struggles have abated, as both sides have matured in their understanding of the problems. However, public opposition can still be a potent force. Just recently an American public utility filed for bankruptcy, the first ever, because of the impossibility of bringing its controversial nuclear power plant onstream. When the relevant public is really aroused, proposers and regulators face a most difficult situation. As the American Harvey Brooks has said, it is scientifically impossible to prove an impossibility; and this may seem to be what is demanded by protestors. Equally, it is politically impossible to prove an 'acceptability' of a risk, when a public sees it being imposed on them by an organisation that lacks human concern and perhaps also technical competence. Hence the new politics of NIMBY (Not In My Back Yard) is an important factor in any technological or industrial development.

It is clear that a sober and responsible mood now prevails among those responsible for the current work of managing the risks of genetic technologies. It is therefore possible for there to arise a new and mutually beneficial dialogue between the scientific and regulatory experts on the one hand and the representatives of the public interest on the other. It is now understood that the general public do not attempt to assess the risks being imposed on them, but rather assess those who either create or regulate them. In this they are not being irrational: just like a lay jury (and a lay judge) in our judicial system, they evaluate evidence to a great extent from the quality of the testimony of witnesses. Of course they can be mistaken, but the policy is prudent and effective both in practice and in principle. To continue the juridicial analogy, all it needs is for safety to be respected and to be seen to be respected, and the problem of public confidence can be resolved.

A genuine policy of this sort brings its own benefits for the work of regulation and control. For if those in charge see it merely as a cosmetic exercise, placating or appeasing an ignorant public, then the work will eventually deteriorate. The prevention of accidents of any sort requires continuing commitment and morale at all levels, especially from the top down. Our culture and personal attitudes are not tuned to the sort of success that is measured by nothing happening. The best planned system of safety will atrophy and decay without regular injections of commitment. The natural consequence of a good safety record is complacency. Equally naturally this would be followed by corruption, as realized in the concealment of incidents that might spoil the good record. If those in charge of a safety programme believe that their scientific knowledge or personal integrity makes their system immune from danger, then the consequences of their attitude will manifest sooner or later, causing harm to all concerned.

What is needed, then, is a certain attitude of scientific humility, an awareness of partial ignorance and of the human fallibility of those operating the safety system. With this would go an appreciation of the role of independent observers and even critics, in keeping the work up to a good standard. Conversely, people who are experienced in public affairs would from their side recognise the genuinely good intentions, and the real special competences, of those managing and regulating the system. There could then be a constructive dialogue (not necessarily always smooth) rather than a destructive polarized debate. There are precedents for this approach from the 1970s. One was in Cambridge, Massachusetts, after the shouting died down. A Committee of citizens representing a cross-section of the population reviewed the problems at Harvard University, and eventually accepted regulations which were only a slight modification of those originally proposed. The other was the original GMAG, where representatives of employees and of the public interest were equal members with the scientists, and (behind closed though leaky doors) engaged in a very effective common dialogue on the construction and operation of the regulatory system.

A recent example from the European Community shows how the different perspectives of scientists and public-interest representatives can be complementary in the full appreciation of a hazards problem. The case was the 'Lamming Report' on the use of certain steroid hormones for farm animals which judged that the substances in question were acceptable under certain conditions; and yet the recommendation for adoption was rejected by the European Commission. The first reaction to this was that in such emotive cases, 'political realities outweigh scientific facts'. Although this may have been intended as a cynical comment on the regulatory game, there was a kernel of important truth in it, once we interpret 'political' in a sufficiently broad way. For (as Wynne has shown in Chapter 3) those who opposed the report did so on the basis of its explicit assumption that good practice would prevail in the use of the substances. Such an assumption is natural for the scientists, since their expertise is not in the field or farmyard, and perhaps even more so for the Commission staff who prepared the report. But for those with such experience, the assumption was both crucial and highly questionable; and it was on the basis of the realities of an imperfect world of farmers and their employees that they objected. In any discussion of a proposed diffusion of a new substance, there is a need for both perspectives, the scientific and the practical, and perhaps next time the scientists will be more aware of this other dimension of the regulatory task.

The effectiveness of a proper attitude is confirmed by recent experience of planned releases. Where public concerns have been respected, as in the United Kingdom, there have been no problems; elsewhere, the record is mixed. But again, the respect for the public must be genuine, or else sooner or later the deception will become patent, with harmful consequences for all. This regulatory problem may not be completely in the 'trans-science' category, but it is a long way from the laboratory research that created it.

The skilled management of uncertainties

So far the discussion, and recommendations, have been cast in rather general terms, concerned with the sorts of skills and attitudes that would be appropriate for effective work in the assessment and regulation of the hazards of genetic technology. Are there more specific points, recommendations that can be translated into ordinary practice? Clearly, the assessment of the hazards is a task that should be made amenable to standard, routine procedures. Otherwise we would always be dependent on personal judgements that by their nature are not suited for detailed public criticism and dialogue.

'Risk assessment' is a well-developed technique for studying the hazards of physical and chemical technologies. It has something to offer to this case, but the assumptions underlying its operations prevent it from being the whole answer to our problems. For, as so far developed, it is quantitative. Every stage in a sequential process must be assigned a numerical probability. The final assessment is derived from a manipulation, simple or complex as the case demands, of those numbers. Where there are elements or blocks of the sequence for which there are no numerical probabilities that can claim realism or even plausibility, the technique has little or no relevance. This is the situation we face, as I made clear in my discussion of 'ecological wisdom'. If we are to manage the severe uncertainties that affect the hazards of genetic technologies, less rigidly quantitative techniques will need to be employed.

Our task is not so much to replace the standard risk analysis, as to enrich it. Instead of probabilities all reduced to a linear quantitative measure, we should examine the structure of our uncertainties. On the basis of such an analysis, we will be able to provide guidance to our judgements which, although not cast in a precise numerical form, may be equally effective for the practice of assessment and regulation. There are two complementary approaches to the management of the uncertainties we encounter: a process of elicitation of their structure; and the codification of its results.

For the elicitation of uncertainties, we recall what happens in real scientific debates, as opposed to the simplified models of them imagined by philosophers. Every argument is a complex structure of evidence and inference; evidence is never perfectly true and inferences never perfectly conclusive. So debate is conducted on lines much closer to those in, say, history, than in the explication of logical structures. The credentials of each element or link are scrutinized; and the criticism or defence of them introduces more evidence and inferences, which themselves are controverted. There is no limit in principle to the number of iterations of this process, or its complexity of interweaving of issues. This is why scientific debates usually cannot be tidily resolved, and why, in the absence of new evidence universally accepted as practically conclusive, a debate can sputter on until resolved by exhaustion or ennui.

When a field is less matured and more controversial, the debates can become even more involved and inconclusive. In the case of policy-related research, there are difficulties of another order. These are quite familiar to those working in sensitive areas of toxicology and epidemiology, although they are not the sort of thing with which laboratory-based scientists must normally cope. For example, because of the institutional location of many persons who might be providing information, it cannot be automatically assumed to be the product of disinterested research of the classic sort. Also, much of the information on which the relevant evidence is based, is not in the public domain normally taken for granted in laboratory research (although perhaps less so now in fields close to commercial application). Obtaining such information may involve struggles with administrators and politicians and can even encounter difficulties under a variety of laws.

Hence, although most information needed for the assessment of the hazards of biotechnology can be hoped to be of a more straightforward scientific sort, those who do the work must be prepared for difficulties of every sort, cognitive and social. It would be impossible to encompass all these possibilities and complexities into a single scheme, but a natural ordering of relevant questions can be accomplished so that enquiry and debate can have some coherent structure. For a common example of a possible procedure suppose we are told, 'There is no positive evidence that . . .' — applying usually to some harmful effect. This is usually understood as being equivalent to, 'The known evidence is negative that . . .'. An obvious inference is made: science know it's safe. However, recalling the case of the cancers at the Institut Pasteur, we are aware that the absence of positive evidence is not at all the same as the presence of negative evidence. So our first response, as aware persons in this field, would be, 'Has the evidence been searched for, or researched, in the public domain?' If not, then the logical difference between the two negative statements is apparent.

A further interrogatory sequence would pursue the questions of why the evidence is absent, whether it can be obtained within the time-frame of the current policy process and if not what should be done in light of the relevant ignorance on the problem. A similar range of questions can be presented in the case of information which is of inadequate or dubious quality. All such questions elicit the sorts of blocks to our knowing the information, be they of accident, unfeasibility, impossibility or prohibition. They also lead naturally to the questions of how the deficiency might be remedied, and if not, the consequences of our continued lack of adequate knowledge, and hence the sorts of decisions that must be made, on substance and procedure, in those circumstances. A full display of such an elicitation might be best done with an 'expert system' on a computer. This might appear to be an overly elaborate solution to a problem that is normally solved informally by common sense, but given the intricacy of the structure of the sequence of conditional questions, it is worth considering as an experiment.

Finally, we come to the task of designing a representation of our uncertainties. Chapter seven in this volume outlines a notational scheme which has been developed with a colleague, Silvio Funtowicz. For our present purposes, the most important element of the scheme is the 'pedigree' category, which has a code representing an evaluative account of the information, either in terms of the history of its production or of its intended use. For each sort of information, an appropriate pedigree matrix is designed. It can be used both for representing the results of an evaluation, and also for eliciting that evaluation from experts in the field. It would be a challenge, interesting and worthwhile, to prepare a pedigree code for the information relevant to the hazards of genetic technology.

Conclusion

The hazards of genetic technology are far from being the most urgent environmental problem facing humanity at the present time. Yet there is as much potential for benefit and for harm from this technology over the next half-century as from any other.

Since both the technology and its regulatory effort are both still in an embryonic state, there is no question of our identifying 'clear and present dangers' which require immediate attention and action. Rather, we are trying to establish the framework of the problems, to see what will be involved in maintaining a successful regulatory effort and a co-operative relationship between industry and the public, as the pace of innovation and diffusion grows. We have concentrated on the uncertainties affecting the assessment and regulation of genetic technologies since these seem to be more severe and complex here than in many other important areas. We have tried to show how these uncertainties, however many problems they cause, are no occasion for abandoning a scientific approach to the problems. Rather, developments within science, as it meets the challenges of an environmentally conscious age, should be adequate for an effective and responsible management of these new hazards while ensuring that the benefits of this new technology will be realised for the common good.

Public awareness of biotechnology

Forewarning

There are a number of difficulties apparent when trying to chart the public's understanding of biotechnology. Foremost amongst these is that biotechnology is difficult to define. In very general terms it is the application of scientific and engineering principles to processes involved in the manipulation of biological agents to provide goods and services (Davies, 1986). However, this is not a very precise definition and as Yoxen (1987) notes, biotechnology is essentially multidisciplinary involving the interaction of a number of fields of scientific endeavour including enzymology, bioelectrochemistry, plant cell biochemistry and industrial microbiology. Yoxen suggests that the two areas most often associated with this field in the public's mind are human gene therapy and invitro fertilisation, yet arguably these are not part of biotechnology.

There is also a good deal of uncertainty attached to the potential harmful effects of genetic manipulation on the physical environment let alone the social environment and these uncertainties present further difficulties. Measurement of the public's understanding of the risks and benefits of biotechnology in general, and specific techniques in particular, is difficult enough and is compounded by the disagreements among the scientific community.

A starting point for the investigation of public attitudes towards biotechnology might be a justification for bothering at all. Given the uncertainties in this area why not leave the field to the respective scientific experts to continue their researches within the established methodological and regulatory controls. There are a number of reasons why such a proposition is unsatisfactory. First, as Covello *et al* (1986) point out there is a democratic requirement to alert the public about possible risks to their health or damage to the environment. Second and more pragmatically, as Cvetkovitch *et al* (1987) note, the desire to communicate with the public is not only an appropriate response to the citizens' right to know but also establishes a relationship between expert and public which is useful when it is necessary to give information in the aftermath of a disaster. Third, at a more general level the Royal Society argues for better public understanding of science (The Royal Society, 1985). Their report suggests that hostility and indifference to science, which appears to be more common in Britain than in other countries, 'weakens the nation's industry' p. 9) and ultimately national prosperity. Finally, there is another important issue: that of public trust in the scientific community and managing institutions. Evidence from opinion polls (MORI, 1985) suggests public confidence in these is not very high.

The preceding section discussed principles of scientific methodology and legislative regulation to obviate the dangers of genetic manipulation. This part looks at the limited available evidence on public attitudes towards biotechnology and offers some suggestions to enhance public understanding. Attitudes encompass what people know about a particular object, how they feel about it and their behavioural intentions towards it. For the developers and regulators of biotechnology it is of importance to know whether there will be adverse public reactions and non-acceptance of techniques or processes. The implications of this touch upon issues of advocacy of the risks and benefits of biotechnology and the manner in which the public are informed. Lee (1986) discusses this with respect to communication of information about chemical hazards. He differentiates between persuasive communication and propaganda. The latter involves the deliberate distortion of information and possible imposition of danger on an unsuspecting local community. Lee concludes (p. 152), 'if a government agency believes that ionizing radiation or vaccination or a hazardous waste repository are reasonably safe according to the norms of our society, then they have a clear duty to actively persuade their constituency that this is the case'. The same may be said to hold for biotechnology. Yoxen (1987, 90) proposes that for biotechnology 'a policy of openness by regulating authorities and intending experimenters is essential if public disquiet is to be stilled and the underlying scientific assumptions laid bare for a proper debate'.

Developmental stages of technical innovation

The processes whereby scientific information is first generated by the research community, then developed for application and finally filters through to society as a whole are represented as a developmental sequence by Mulkay (1972). Jacobsson *et al* (1988) point out that for twentieth-century technologies such as the nuclear, electronic and now microbial the time for maturation has grown progressively shorter. In 1981 there were no commercial applications of a genetically engineering process. By 1983 firms were actually using recombinant DNA technology on an industrial scale. Yoxen (1987, 15) states:

> Both the internal developmental mechanics of biotechnology and the influences of the external, commercial and political environment will affect its rate of growth and its eventual form. Technologies embody cumulated social and economic choices. Their evolution depends upon which past choices have been made and which future choices are seen as desirable.

Jacobsson *et al* (1988) propose that it is important to be able to predict the 'hot spots' of scientific advances. Indicators might include the analysis of patents, auditing of available products, examining the wider or related scientific fields. They suggest however that social and political influences are more difficult to predict. A contribution towards understanding these influences has been made from a series of studies conducted at the University of Surrey. These investigators examined the controversies surrounding the siting of nuclear power plant and radioactive waste repositories. The development of public awareness emerged through a number of stages (Brown, 1986). The stages are somewhat overlapping and it is not suggested they occur in a particular sequential order. However it is evident that local complaints and local actions occur early on in the development controversy. A critical mass of these and/or a major accident arouses national media interest. Public confrontation or litigation may also occur and feature as news stories. Concurrent with these developments, special interest groups may take action and actively campaign, also drawing the media's attention. Often following accidents and/or media pressure there is a hastening or adjustment to safety regulation. Finally the issue may become the focus of party political consideration. In the case of nuclear energy, the first thirty years of its development were relatively non-problematic from a public acceptance point of view. Planning permission for nuclear power stations in the early days were comparatively short, as the public debate concerning nuclear energy increased in intensity, the Windscale Inquiry and then the Sizewell Inquiry established new records for the length of time they took to deliberate.

Gradually, there evolved local unease and local complaints around plant. In 1957 there was a serious accident at the Sellafield Nuclear Reprocessing Plant. At the time, the Sellafield fire did not around widespread concern and remedial reaction was confined to the local area. However, since that time, groups, such as Friends of the Earth, focused attention on the dangers posed by nuclear plant. They utilised the media to draw attention to their case. Following the plant inquiry into the Thermal Oxide Reprocessing Plant (THORP) at Sellafield in 1977, Friends of the Earth averaged 300 mentions per month in the national press (Ward, 1983). Public opposition to siting nuclear plant or radioactive waste repositories has grown and become increasingly confrontational. In 1982 the Central Electricity Generating Board took Cornish protesters to court to prohibit them from interfering with test drilling. Films, such as the *China Syndrome* or *Silkwood*, and more recently a BBC serial, *Edge of Darkness*, highlight the issues in the popular media reaching beyond the purely sectional interests of the environmental activists. Accidents such as Three Mile Island in 1979 and Chernobyl in 1986 have aroused greater, more general concern which exacerbate the growing opposition to nuclear energy. Finally, the issue has become a party political matter and has made its appearance in the manifestos of Britain's political parties.

Emergence of public concern about biotechnology

As yet, there has been only a modest level of expressed public concern related to biotechnology as measured by activities of local groups, opinion leaders or the media. There has not been a major catastrophe associated with biotechnology and it has not featured significantly in party political considerations. Thus the public dimension of the potental impacts of biotechnology is still in an embryonic state. Taking opinion leaders first, Pope John-Paul II (1983) wrote a condemnation of experimental manipulations of the human embryo, although he did cite as appropriate research that which explores genetic or chromosome disease and increases of food products. The American, Jeremy Rifkind, leader of a coalition of consumer and environmental groups, has undertaken a great deal of successful lobbying activity to stop the release of genetically engineered micro-organisms into the environment in the United States.

A survey of attitudes of opinion leaders towards aspects of biotechnology was undertaken by Miller (1985) in the United States. This used a methodology in which a purposive sample of opinion leaders from science policy, church and environmental groups participated in a telephone survey. They were asked, 'when people speak about the risks of genetic engineering, what kinds of risks come to mind?' Biological warfare was mentioned by 11 per cent of biologically trained science policy opinion leaders, 5 per cent of the environmental and 4 per cent of the church leaders. Interestingly, whereas 21 per cent of the church leaders thought they were well informed on the issues only 6 per cent of environmentalists took this view. In terms of problems surrounding recombinant DNA research 17 per cent of the science policy opinion leaders thought this a source of controversy, compared to 42 per cent of the church leaders.

The role of the media in drawing attention to biotechnology may be measured by reference to the British Library Association quarterly index of British periodicals and quality newspapers called the British Humanities Index. There were no articles indexed under the heading biotechnology until 1980 when nine appeared. Since then there has been an average of ten articles indexed per year. There has also been a modest rise in the number of articles indexed under the heading genetics or genetic engineering. Most of the articles draw attention to the darker aspects of genetic manipulation. Examples of headlines include 'Pandora's Box of Genes' (*Economist*, 5 March 1977), 'Diary with Doomsday' (*Spectator*, 10 June 1978), 'Will the Gene Hunters lead us into a Nightmare World' (*Sunday Times*, 8 January 1984), 'Keeping Cells Safely under Lock and Key' (*Guardian*, 10 January 1985), 'Just who will cash in on the brave new world of life forms created purely for profit' (*Guardian*, 25 April 1987). The role of the media in influencing attitudes towards technology, notably nuclear energy, has been the subject of research by Slovic and his colleagues. They argue (Slovic *et al* 1982) that in judging the riskiness of events or processes the public rely on memory or imagination which in turn are affected by factors such as vividness or emotional impact. Recall is influenced by the so-called availability heuristic or bias in beliefs about the probability of occurrence of disasters because of their familiarity. Risks from nuclear power are particularly prone to exaggeration because of the intense media coverage and its association with the vivid and imaginable dangers of nuclear war. The examples of headlines relating to biotechnology given above demonstrate associations with images drawn from Huxley and Orwell. The metaphors are powerful and telling, although it remains to be seen the influence this kind of coverage will have on shaping attitudes.

Another indicator of the salience of biotechnology as an issue of public attention is the degree to which special interest groups take up the issue. As has already been mentioned, the consequences of biotechnology have become a focal environmental concern in the United States. At present this does not appear to be so in the United Kingdom. Yoxen (1987) indicates environmental groups in Britain cautiously welcomed the proposal to release a recombinant virus in order to test a possible strategy for killing caterpillars that destroyed pine trees. Their argument was that such techniques were likely to lead to a reduction in the use of chemical pesticides. Erlichman in an article appearing in the *Guardian* (25 April 1987) draws attention to British environmentalists' activities concerning biotechnology in the following terms, 'what is worse, environmentalists in the UK have not even woken up to the dangers'. Mr Rifkin successfully blocked researchers who tried to release genetically engineered plants and organisms in the United States. When a genetically engineered virus was released last year by British scientists at the Natural Environment Research Council Institute of Virology, pressure groups such as Friends of the Earth raised hardly a whimper'.

Trade unions have addressed the issue of risks of biotechnology to some degree. Their involvement is largely preoccupied with safeguarding the interests of their members and their activities confined to representation on advisory committees.

The degree to which issues relating to biotechnology have impinged on the public at large can be measured by surveys of public attitudes. Much of this research is concerned with attitudes towards new technology or science and technology in general rather than biotechnology *per se*. MORI (1985) undertook a survey on behalf of the Technical Change Centre. A national sample of British critizens (N = 1,824) were asked a series of questions including the following: 'Please read out [from a supplied list] all the terms you think you know the meaning of'. Most, 85 per cent, mentioned nuclear power, 78 per cent computer program, 52 per cent genetic engineering, 39 per cent biochemical engineering, 32 per cent biotechnology, 28 per cent enzyme technology and fewest, 22 per cent selected in vitro fertilization. However the survey did not ask respondents to define these terms and it is possible they were answering by way of recognition rather than any knowledge of their meaning. The survey did reveal some interesting differences, e.g. 78 per cent endorsed the development of new medical techniques, but 63 per cent were against the development of new processes to produce synthetic food and 68 per cent were opposed to the cross-breeding of animals.

Consumer attitudes towards technological innovations in food processing were explored by Lee and his colleagues (Lee *et al* 1985; Lee 1986). Here in-depth interviews were conducted with 40 respondents in which qualitative responses were elicited. Moral judgements were frequently invoked against genetic engineering manipulations of plant or animal products. Typical comments included 'it's unnatural', 'I don't think you should mix science with food'. The same investigators conducted a quantitative survey of 203 individuals in which 70 per cent indicated they thought genetic engineering is morally wrong, 62 per cent that it is unnatural and 27 per cent found it frightening. In this survey 42 per cent thought interspecies breeding wrong and 40 per cent believed it to be unnatural.

In various surveys conducted in the Department of Psychology, University of Surrey, from 1982 to 1986, a general question was posed in order to elicit those issues that the public volunteered as causing either themselves personal worry or were of concern to society as a whole. Biotechnology did not feature as a volunteered personal or society worry or concern. Most concern or worry related to unemployment or the threat of violence (Brown 1986).

Conclusion and recommendations

As yet, in the United Kingdom at least, the issues surrounding the risks and benefits of biotechnology have not appeared on the agenda either of special interest groups or the media in a concerted or regular manner. What available evidence there is about public opinion indicates that there is little awareness of or consideration given to biotechnology. However, it would be complacent to assume this lack of manifest interest or criticism means tacit consent to the speed or type of developments in biotechnology.

The survey by Miller (1985) indicates there is a demand for information about biotechnology. The style and content of information and the mode of delivery needs careful consideration. In the United States there have been a number of innovatory experiments in public participation in debates about the risks and benefits of recombinant DNA. Seeley and Wolpert (1982) described the experience of a science court conceived in Cambridge, Massachusetts comprising of a lay jury hearing evidence about DNA research and discussed earlier in this chapter. Given an appropriate setting, the lay public were able to cope with technical material about complex scientific issues. However, biotechnology involves considerations of moral and ethical matters as well as scientific assessment of risks and benefits. There is evidence for some public concern about these moral aspects, as well as biological consequences of releasing genetically engineered micro-organisms into the environment. Biotechnologists, regulators and developers are already and will continue to be under public scrutiny. The example of nuclear energy has taught us that there is a demand not only for information disseminated through the mass media to inform the general public about the issues but also a need to reassure them about the credibility of those who are responsible for safeguarding health and the environment. This involves making the process of accountability more open and the watchdog or advisory committees more accessible to lay membership.

Accountability and dissemination of information about nuclear power have been largely reactive and often of the damage limitation variety. Dialogue with the public over biotechnology has the opportunity to develop proactively. Thus there is time to evolve an imaginative programme of discussion and dissemination of information that fulfils the needs of the uninquiring as well as interested public. Tailoring the requirements of such a programme requires on the one hand an understanding of the cognitive structures of knowledge and beliefs that people have available to them and how people with different cognitive systems assimilate and incorporate new knowledge. Research is currently underway inspired by an ESRC initiative into public understanding of science. On the other hand innovative mechanisms such as the science courts might be developed to provide more flexible methods to involve the lay public in society's consideration of pertinent issues. There is a challenge too in reaching the emerging citizen whose life may be profoundly affected by developments in biotechnology. It is thus important to introduce children and young people to the ideas and concepts of biotechnology. Science and technology museums such as the highly successful San Francisco Exploratium Science Center and the British Exploratory in Bristol offer bold new ways of involving children in learning about science and technology through hands-on experience. As yet the biological sciences are not well represented and there is scope to develop ideas of presentation within these new-style museum experiences.

There are no instant corrective prescriptions for designing information or public relations' packages to create either understanding or acceptance of biotechnological innovations. However, there is time available to give consideration to the issue of public understanding of and participation in scientific developments that will impinge on their medical care, food and energy production or treatment of waste products in a way that avoids distrust and hostility towards experts as experienced by the nuclear industry.

References

Brown, J., 1986, *Public Perception of Risks: The Case of Dioxin*, Royal Society Study Group, London.

Colingridge, D., 1980, *The Social Control of Technology*, Frances Pinter, London.

Covello, V., Winterfeldt, D. von, and Slovic, P., 1986, 'Risk Communications: A Review of the Literature', *Risk Abstracts*, **3**, 171–82.

Cvetkovitch, G., Vlek, C. and Earle, T., 1987, 'Designing Public Hazard Communication Programs about Large Scale Technologies', Batelle Institute, Seattle, mimeo.

Davies, D.K. (ed.), 1986, *Industrial Biotechnology for Europe; Issues for Public Policy*, Centre for European Policy Studies, Commission of the European Communities.

Jacobsson, S., Famison, A., and Rothman, A., 1988, *The Biotechnological Challenge*, Cambridge University Press, Cambridge.

Lee, T.R., 1986, 'Effective Communication of Information about Chemical Hazards', *Science of the Total Environment*, **51**, 144–83.

Lee, T.R., Cody, C., and Plastow, E., 1985, *Consumer Attitudes Towards Technological Innovation in Food Processing*, University of Surrey, Guildford.

Library Association, 1977–87, British Humanities Index, L.A., London.

Market Opinion Research International (MORI), 1985, *Public Attitudes to New Technology*, MORI, London.

Miller, J.D., 1985, 'The Attitudes of Religious, Environmental and Science Policy Leaders towards Biotechnology', *Recombinant DNA Technical Bulletin*, **8**, 141–64.

Mulkay, K.J., 1972, *The Social Process of Innovation*, Macmillan, London.

Pope John-Paul II, 1983, 'Ad eos qui conventiu de biologae experimentis in factican civitate habito interfuere', *Acta Apostolicae*, **LXXV**, 35–9.

Royal Society, 1985, *Public Understanding of Science*, Royal Society, London.

Seley, J.E. and Wolpert, J., 1982, 'Negotiating Urban Risk', in Herbert, D.T. and Johnston, R.J. (eds), *Geography and the Urban Development; Progressive Research and Application*, Wiley, Chichester, 279–305.

Slovic, P., Fischhoff, B. and Lichtenstein, S., 1982, 'Psychological Aspects of Risk Perception', in Sills, D, Wolf, C.D. and Shelanski, V. (eds), *Accident at Three Mile Island; The Human Dimension*, Westview, Boulder.

Ward, H., 1983. 'The Anti-nuclear Lobby' in Marsh, D. (ed.), *Pressure Politics; Interest Groups in Britain*. Junction Books, London.

Yoxen, E., 1987. 'Social Impact of Biotechnology on Living and Working Conditions', European Foundation for the Improvement of Living and Working Conditions, Shankill, Co. Dublin.

Chapter 6

The Plural Rationality Approach[1]

Peter James and Michael Thompson

Like many social scientists, we attribute public unwillingness to accept certain low-probability risks not to 'irrationality' but rather to the everyday application of qualitative as well as quantitative criteria. Just as most people buying a video or hi-fi do not base their decision solely on price or performance data but also consider such intangibles as design and the reputation of the manufacturer, individuals faced with a given risk take into account not merely the probability of harm but also the credibility and trustworthiness of whoever generates the information and manages the safety and other precautions which ensure its accuracy.

To analyse these qualitative dimensions we utilise the grid/group methodology developed by the cultural anthropologist Mary Douglas, and her co-workers (Douglas 1982a). Douglas identifies four basic forms of social organisation and four associated 'thought worlds' which serve to legitimise and reproduce them. Differences in concepts of human cognition, time and many other variables between the different thought worlds are so profound and incommensurable that they can be regarded as constituting different kinds of rationality. The acceptability of risks is partially determined by the extent to which they are perceived as supporting or threatening these different, and contending, ways of organising.

Four rationalities

Like other holistic social theories, grid/group analysis posits that individual thought and behaviour are shaped more by social influences than personal volition (Douglas 1982). It uses two indices to reduce the manifest variety of social influence to manageable proportions (Douglas 1982a,1982b; Gross and Rayner 1985). (See Figure 6.1.)

The group index measures the extent to which individuals belong to bounded social units. Those who do belong (high group) will tend to have frequent interactions with other group members, make strong distinctions between their group (us) and the outside world (them) and be strongly influenced by a collective ethos (for they may otherwise be categorised as deviant). Conversely, unincorporated individuals will have more extended

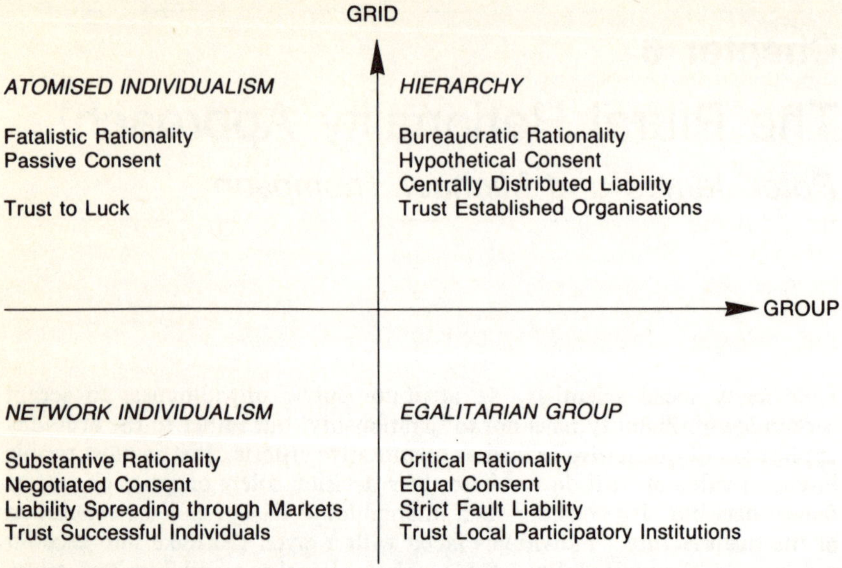

Figure 6.1 Four types of social organisation

Adapted from Rayner (1985)

and open-ended social networks, weaker ties with each other and be less subject to group pressures on thought. The *grid* index is a composite measure of social constraints on individual behaviour. These constraints may be derived from group membership but can also be externally imposed, for example by gender roles, status differentials, kinship obligations or racial or religious restrictions. High-grid individuals are subject to a great many restraints and therefore have access to a limited range of socially acceptable behaviour. Low-grid individuals are relatively free to behave as they want to.

All these categories have been previously identified by social theorists, but never delineated in their entirety nor in a systemic manner. Weber and others have noted the differences in thought and behaviour between hierarchical societies and the individualistic networks that are characteristic of markets. Hierarchies are made up of bounded social groups, each of which is in an orderly and ranked relationship with each other.[2] The need to maintain this internal order gives rise to a bureaucratic rationality that stresses the importance of correct procedure and is more concerned with the proprieties of who does what than with trying to evaluate the outcome (if there is one). Market societies stress the autonomy of individuals and their resulting freedom to bid and bargain with each other: they have a substantive rationality — the 'bottom line' is what they care for, not the relational niceties of the people who happen to have come together to achieve that result.

But this is an inadequate social taxonomy. Many individuals reject both the competitiveness of the market and the inequalities of the hierarchy: they prefer the co-operative fraternity (or sorority) of voluntary egalitarian groups. The egalitarian ideal is always under threat from internal manifestations of the individualism and hierarchy which its followers oppose in the outside world: they must therefore have a watchful, critical rationality. It is also a cruel travesty to describe all those who are individualised as busy, networking entrepreneurs. Some face so many prescriptions on their behaviour that they have minimal freedom of choice, for example, the unemployed, trailing from one welfare centre to another *ad infinitum*. These atomised individuals are on the margins of society and their inability to influence events this way or that engenders a fatalistic rationality in which outcomes, good or bad, are simply to be enjoyed or endured, but never achieved.

Of course, these categories are dynamic, for individuals are always moving from one to another (and therefore, to some extent, choosing their destiny). They are also ideal types that are, in practice, modified by other forms of social classification such as class, gender and so on. Nonetheless we believe that the four categories are enduring, protean structures which influence all social beings.

Risk acceptance

The best-known application of grid/group theory to the subject of risk is the book by Mary Douglas and Aaron Wildavsky, *Risk and Culture*. Unfortunately, this is flawed by its partisan approach to environmentalism and its condensation of the four rationalities into an overly simplistic centre/border dichotomy and has sometimes been more of a hindrance than a help in demonstrating the validity and objectivity of the authors' methodology. The work of Mary Douglas (1986b) alone, Douglas MacLean (1986), Steve Rayner (1984, 1985) and Michael Thompson (1980) provides a more considered grid/group approach. MacLean and Rayner (on whose work the following section is based), identify the critical factor in risk acceptance as being the extent to which the procedures of consent and the distribution of risk liabilities conform to community norms of fairness.

Modern hierarchies assume a social contract between themselves and their members, with the latter gaining access to selective benefits in exchange for subordination to collective authority: an arrangement which can be termed *hypothetical consent* (Rawls 1971; Rayner 1984). Risk liabilities are therefore distributed by a central bureaucracy. Provided that correct procedures are followed — defensible methods of assessing and managing risks by expert, enduring and therefore trustworthy organisations — consent can be taken for granted, even if outcomes are unequal. The need to sacrifice individual wishes for the good of the whole (group) is a constant refrain of hierarchical life and it is therefore unsurprising that hierarchies are so puzzled and angry when the correct procedures are followed and recalcitrant minorities still refuse to accept a particular risk.

They do so because risk decisions achieved through hypothetical consent violate the consent principles of network individuals and egalitarian-bounded groups. In a world of entrepreneurs, consent must be negotiated between individuals, who are considered far more trustworthy than faceless organisations. Risk liabilities are therefore spread by market mechanisms (or simulations of them) and, because individual preferences and utilities are deemed to be unique, are unevenly distributed (Calabresi 1970; Rayner 1984).

Egalitarian groups stress the importance of voluntary and equal consent to risk — conditions which cannot be met by the subordination of hierarchy or the lack of resources of many participants on markets. This model of consent presupposes an equitable distribution of risk liabilities based, as Rayner has argued, on a strict fault principle of liability (Calabresi 1970; Rayner 1984). Egalitarians trust neither coercive organisations nor egoistic personal networks to achieve this: they have faith only in local, participatory institutions.

Atomised individuals are largely ignored in risk debates. They do not expect the world to be fair and are unsurprised that both bureaucratically distributed and market spread risk liabilities end up on their doorsteps. They know that human institutions will always fail them and trust only to luck. Their consent is passive, dependent on minimal requirements being met and lip service being paid to their sense of dignity. Only when these elementary needs are denied do they form the networks or join the groups that would change their position within the social order.

Plural rationalities and nuclear power

There are many factors in public hostility to the nuclear industry but prominent among them are its high degrees of bureaucratisation, its remoteness and uncontrollability, its stress on expert authority and a perceived lack of choice and absence of open decision-making (Lee 1987). These concerns can be restated as a violation of the consent principles of individualistic and egalitarian ways of life and also of the liability principles of the latter. As a result, their members doubt the fairness of nuclear decisions and consequently mistrust the institutions which assess and manage nuclear risks.

Network individuals have problems with a technology that so relentlessly controls and channels individual initiative and gives so little scope for personal management of risk. However, it is important to note these misgivings are not ones of principle for entrepreneurs are archetypal NIMBYs (not in my backyard). Their limited social (and temporal) horizons make them indifferent to nuclear facilities being sited in someone else's backyard: they may even have it in their own if they can gain some personal advantage, for example, cheap electricity or tax concessions. (By contrast, if local circumstances force hierarchical bodies only to oppose nuclear installations, Rayner (1984) predicts they will be bought off by community benefits that can be distributed by established institutions.

Egalitarian consent principles are more fundamentally violated by nuclear power. It is difficult to achieve any kind of local control over a nuclear installation. Moreover, as O'Riordan and many others have observed, nuclear power has been identified as a central element in modern industrial society and therefore come to symbolise the material growth (and consequent inequalities of wealth), centralised power and expert authority that are so corrosive of the egalitarian ideal. Egalitarians are NIABYs (not in any backyard) and are opposed to the technology in principle.

This analysis highlights the importance of scale as a major influence on the acceptability of risks for, the greater the scale, the more the consent principles of egalitarianism and individualism are violated and the more likely it is that the trustworthiness of nuclear institutions will be questioned. This is true of the scale of the plant (where size increases create a greater intrusion into both the physical and cultural landscape), the scale of the worst credible accident (with feelings of dread and helplessness increasing in line with the number of likely deaths) and the scale of the national and international commitment to nuclear power (for, the greater the commitment, the more people feel that choice is reduced). In this respect it is interesting to note Alvin Weinberg's theory that societies have been willing to bear only a constant amount of risk from nuclear power and have therefore demanded stricter safety requirements for individual plants as the total number has increased.

If nuclear power is to survive in a hostile world (and one more accident would almost certainly stop further development in most industrial countries) it may need to deal with the scale issue. This is likely to require the accelerated development of small/medium scale 'inherently safe' technologies, an eschewal of large, long-term nuclear programmes and greater efforts on the part of energy decision-makers to ensure (and to be seen to ensure) that nuclear power is only one element in long-term energy development and is not reducing effective choices by, for example, taking a large proportion of energy R and D expenditures or scientific and technical talent.

Interestingly, there is one country — Sweden — which is moving in this direction. Despite being, in per capita terms, one of the most nuclear dependent countries in the world, it has nonetheless decided gradually to phase out its nuclear reactors. Time will tell whether a cold, fossil-fuel-deficient nation can manage without nuclear energy at all; if not, it can be argued that Sweden in the 1990s might be surprisingly well placed to initiate a new stage in the development of nuclear technology. Certainly it will then be difficult for nuclear opponents to argue that alternative sources of energy have not been seriously considered. The blocking of the conventional nuclear route also means that Sweden is taking a leading role in the development of 'inherently safe' reactors and may have the technical and political ability to deploy them early in the next century. The same is true of nuclear waste disposal where, despite the greater strength of its environmental movement, Sweden is far closer to developing a broadly acceptable solution than is the United Kingdom. One reason for this may be the decoupling of waste disposal from further nuclear development, so that potential opponents do not feel that a solution to the waste problem automatically leads to further nuclear expansion (of course, this may be the end result, but what matters is that people do not believe it to be so at present). By thus accommodating to egalitarian and individualist opinion (if the latter exists in so notoriously collectivist a country) Sweden may be buying a breathing space which will allow it to forge a socially tolerated, and therefore more durable, nuclear future than that developing in the United Kingdom and most other industrialised countries.

Plural rationalities and cancer

One of the most striking features of contemporary risk debates is the extent to which they focus on cancer, despite the fact that heart disease is not only a more important cause of death in industrial society but, on average, claims its victims at an earlier age.

Mary Douglas, Susan Sontag and others have shown that diseases become metaphors for social disfunction and therefore taboo subjects. As with syphilis in the early modern era or tuberculosis in the nineteenth century, so it is with cancer today. One reason may be that cancer powerfully symbolises some of the central concerns of our different rationalities. Hierarchies generate order and regularity through strong internal and external boundaries. The genesis and spread of cancer overcomes normal metabolic control mechanisms and can be interpreted as a rebellion of parts against the whole, setting the physical body against itself just as excessive individualism destroys the bonds of community that preserve the social body. The metaphor of cancer as individualism-run-riot has some resonance for group-orientated egalitarians, but far more important is its symbolisation of the internal and external threats to the egalitarian ideal. The ubiquity of carcinogenic agents — in the air we breathe, the water we drink, the food we eat — can serve as a constant reminder of the evil of the industrial world which generates them and thereby used to strengthen community vigilance against the inroads of hierarchy and individualism.

Cancer is perhaps more of a second order threat to network individualism. In this world all disease is anathema because it undermines the belief that individuals can manage risk and control their destiny. Insofar as cancer has become the primary metaphor for loss of control it is perhaps because of its salience to the other rationalities and therefore in public opinion as a whole. One can further speculate that because cancer is so salient to the different rationalities, it has become a general symbol of the cruelty and randomness of life and the inevitability of death that, in a sense, all human culture is seeking to deny through the creation of meaning. Perhaps this is why so many cancer victims speak of shame and rejection: their existence is a reproach to the established order and must therefore be denied or blamed on the victim. Interestingly, one would expect this to be less true of egalitarian groups, for whom cancer can bear witness to the irredeemable evil of the outside world and therefore strengthen their group solidarity.

Finally, it is worth pointing out that — as a latent, probabilistic ailment — one would not expect cancer to have a high salience to fatalistic, atomised individuals. And indeed health educators are forever despairing that their warnings about cigarettes are ignored, or trade unionists that unorganised workers can always be hired for dangerous work. Plural rationalities — and the work of medical sociologists such as Cornwell (1984) — tell us that this is not merely the result of lack of concern or information but rather an entirely appropriate response to a world which limits both collective solidarity and the development of individual self-confidence.

Notes

1. This paper is a modified version of one presented to the UKAEA symposium on Risk Perception and Safety Targets for Major Accidents, organised by the Environmental Risk Assessment Unit, University of East Anglia on 16 October 1987.
2. Grid/group analysis was developed, and applies most specifically, to the relatively simple social units of the pre-modern world. The aggregation of simple units that is characteristic of modern industrial society creates more complex units, particularly the nested or complex hierarchy. This is often sufficiently large in scale to contain semi-egalitarian groups and individual networks within and around it, and to accommodate marginal individuals in its lower reaches (for example, casual workers taken on when the workload warrants it). It is therefore legitimate to speak of such organisations as pluralised grid/group regimes. Contrast, for example, the characteristic mix of hierarchy and egalitarianism in Japanese organisations (and Western armies) with the dominance of hierarchy and entrepreneurialism (old-boy networks, competition for promotion, etc.) in the West. One of the key insights of grid/group analysis is that there are different ways of breaking down or softening up hierarchy.

References

Calabresi, Guido, 1970, *The Cost of Accidents*, Yale University Press, New Haven.
Cornwell, Jocelyn, 1984, *Hard-Earned Lives*, Tavistock, London.
Douglas, Mary (ed.), 1982a, *Essays in the Sociology of Perception*, Routledge & Kegan Paul, London.
Douglas, Mary, 1982b, *In the Active Voice*, Routledge & Kegan Paul, London.
Douglas, Mary, 1986a, *How Institutions Think*, University of Syracuse Press, Syracuse.
Douglas, Mary, 1986b, *Risk Acceptability According to the Social Sciences*, Routledge & Kegan Paul, London.
Douglas, Mary, and Wildavsky, Aaron, 1982, *Risk and Culture*, University of California Press, Berkeley.
Gross, Jonathan and Rayner, Steve, 1985, *Measuring Culture*, Columbia University Press, New York.
Lee, Terence, 1987, 'Risks and Perceptions', *Atom*, November, 34–5.
MacLean, Douglas, 1986, 'Risk and Consent: Philosophical Issues for Centralized Decisions', in MacLean, D. (ed.), *Values at Risk*, Rowman and Allenheld, Totawa.

O'Riordan, Tim, 1983, *Environmentalism*, Pion, London.
Rawls, John, 1971, *A Theory of Justice*, Belknap Press, Cambridge.
Rayner, Steve, 1984, 'Disagreeing about Risk', in Hadden, Susan (ed.), *Risk Analysis, Institutions and Public Policy*, Associated Faculty Press, New York.
Rayner, Steve, 1985, *Redefining Risk: a Cultural Strategy for Risk Management*, Oak Ridge National Laboratory, Oak Ridge, mimeo.
Sontag, Susan, 1979, *Illness as Metaphor*, Allen Lane, London.
Thompson, Michael, 1980, 'Aesthetics of Risk', in Schwing, Richard and Albers, Walter (eds), *Societal Risk Assessment*, Plenum Press, New York.
Weinberg, Alvin, 1985, *Continuing the Nuclear Dialogue*, American Nuclear Society, Grange, Illinois.

Chapter 7
Managing the Uncertainties of Statistical Information

Silvio Funtowicz and Jerry Ravetz

Introduction

There is a long tradition in public affairs which assumes that solutions to policy problems should, and can, be determined by 'the facts' expressed in quantitative form. But such quantitative information, either as particular inputs to decision-making or as general purpose statistics, is itself becoming increasingly problematic and affected by various sorts of uncertainty.

The new challenges of science, in problems of technological risks and environmental issues, cannot be resolved by the simple extension of traditional laboratory-based procedures, for these have fully revealed their inadequacy. In the debates on environmental and occupational hazards, which are bound to increase greatly before they ever abate, popular conceptions of science tend to change drastically from naïve trust to a dangerous disappointment. In debates on large-scale problems, such as engineering projects constituting 'major hazards' or major environmental intrusions, or in the speculative technologies of nuclear armaments, the dividing line between science, self-interest and fantasy becomes very difficult to discern. The traditional methodologies of scientific research offer insufficient protection against the distortions of reason that modern conditions encourage, extending even to our dealings with the world of Nature.

This chapter forms part of a larger philosophical work, in which conceptual analysis is closely tied to scientific practice as realised in particular problems. We are constructing a methodology in which we articulate the complementarity and interaction of knowledge and ignorance, and of quantity and quality, in all scientific information. In the face of uncertainty, we avoid the mistake, common up to now, of attempting to conquer it through formalisation and quantification. Rather, we use classification and rough-quantification to design simple and effective tools for its management. Such methods will not by themselves be sufficient, but they are, we believe, necessary for the solution of the technical and social problems afflicting science in the policy-related fields.

We have developed and tested a notational system NUSAP (Funtowicz and Ravetz 1987a; 1987b) for the expression of quantitative information, one which provides places for each of the judgements describing the different sorts of uncertainty with which every quantitative statement is qualified. By its means, we can improve the understanding, application and quality-control of all technical information (including environmental and industrial risk calculations). The system depicts the complexity of all such information: its production, its communication and its use as an input in the policy process. By its means, quality-control of statistical information will be improved, and it will be less vulnerable to errors and misleading interpretations.

The system's name, NUSAP, is an acronym for the categories Numeral, Unit, Spread, Assessment and Pedigree. The last three places, on the more qualifying side, convey inexactness, unreliability and border with ignorance, respectively. Familiar analogues exist for the first two of these. Spread is a generalisation of 'experimental error' or 'tolerance' or 'variance'. Assessment generalises measures of reliability or unreliability such as 'confidence limits' in statistical testing or 'systematic error' in laboratory science. The last category, pedigree, does not have a precedent in ordinary scientific practice or statistical technique. We define it as an evaluative account of the quantitative information, in the present case one which exhibits the process whereby the quantity was produced. By means of that account, we characterise the state of the art of the production of the quantity. This exhibits the inherent limitations of the knowledge that can be achieved on its basis, and in that sense demarcates the 'border with ignorance' in that case. The first two places, on the 'quantifying' side of the scheme, are numeral and unit respectively, which are close enough to their traditional analogues to need no explanation as yet. Within each of the five categories, or places, we can employ different notations, representing different practical concepts, as in the examples for spread and assessment above. The choice will depend on the intended application of the particular NUSAP expression.

Principles of NUSAP

NUSAP is a system of notations whereby the various sorts of uncertainty contained in all quantitative information may be expressed concisely and also consistently with existing partial notations. It is designed to be applied to any expression that is given in the form of numbers or more generalised notations. By its means, nuances of meaning in quantitative statements can be conveyed clearly and economically; and various aspects of the quality of the quantitative information may also be expressed. For those who are aware of the need to manage the inherent uncertainties in all their quantitative information, NUSAP can be used successfully for education in the appropriate skills. For those who lack such an awareness but are open to learning, the NUSAP approach can be quite effective in enhancing their perceptions.

NUSAP was designed with several criteria in mind. In general, any new mathematical notations must have properties such as the following:

Simplicity, as a set of signs to be written and read;
Naturalness, as an extension of familiar usages;
Coherence, being effective for several sorts of operations (for example, calculating and data-recording);
Flexibility, suited for a variety of functions of communication;
Adaptability, to a variety of fields and contexts; and
Clarity, exhibiting the good practice of traditional usages, and thereby guiding developments in new applications.

In the present context, three more design criteria are given special emphasis. The notation is designed to express three sorts of uncertainty: inexactness of measurement and of representation; unreliability of methods, models and theories; and the border between knowledge and ignorance. It protects against the misleading use of technical information, by preventing the communication of simple quantified expressions isolated from their relevant qualifying properties. It distinguishes between meaning and meaningless quantitative expressions; this can be accomplished by the identification of pseudo-precision, which is usually a manipulation with meaningless digits.

NUSAP is a 'system' because it is not simply a collection of fixed notations. Rather, it is a set of five distinct categories, each of which has a determinate meaning. We shall first review them all briefly, and then explain them individually in detail. Considering the expression as proceeding from left to right, we start with those which are more familiar, the quantifying part of the expression; and conclude with those less familiar, forming the qualifying part of the expression.

The first category, in this order, is *numeral* (N). We use this term rather than 'number' as a reminder of the flexibility of the system. The place can be filled by a whole number (for example, 425), a decimal expansion (for example, 4.25), a fraction (for example, 4¼), a multi-dimensioned quantity (for example, a matrix), or even a qualitative index (for example, 'strong', as opposed to 'medium' or 'weak'). Next is *unit* (U), which can be a compound entry, consisting of 'standard' and 'multiplier'. This can be important for the representation of aggregated quantities, as '\$K', or perhaps '$\$10^{12}$'. The middle category, *spread* (S), generalises the traditional concept of error. Although this is normally expressed in arithmetical form (perhaps by ±, % or Λn, for 'over a logarithmic range of n') there is a strong qualitative element about it. For *spread* cannot (except perhaps when obtained by a statistical calculation) be given at all precisely. It is always an estimate, whose own spread ('spread^2') is not a useful concept because it leads directly to unending recursion and meaninglessness. There is a way of qualifying the spread entry; it can be done by *assessment* (A), the fourth category in the NUSAP system. This may be seen most familiarly as a generalisation of the confidence limits used in statistical practice. Assessment can also be relevant in a context where the problem does not admit of the calculation of confidence limits. A great variety of notations can be deployed here, ranging from percentages to a simple ordinal scale, such as 'high, medium, low'. The means of arriving at an assessment rating are equally various: it may be calculated statistically; it may be obtained by arithmetical operations from a conventional coding of the last category (pedigree); or it may be the result of a personal judgement.

Hitherto the categories have analogies in existing practice, ordinary or statistical; and it is natural to consider NUSAP as simply an extension and ordering of existing notations. But with the pedigree category (P), a novelty is introduced. By pedigree we understand an evaluative account of the quantitative information; in the present case of the production of the quantity. Evaluative accounts do not normally appear as part of mathematical notations; and for this category we have developed special schemes of analysis and representation. So far we have designed pedigrees for 'research' information and for 'statistical' information, this latter sort will be discussed in this chapter (for applications of the 'research' pedigree, see Macgill (1987), and Macgill and Funtowicz (1988). The pedigree is designed around the particular sort of information under consideration, in the context of its intended use. On the basis of the pedigree ratings, there can be an assessment of the reliability or unreliability of the information, perhaps depending on the purposes of the users. This summary evaluation may be recorded in the assessment place of NUSAP.

NUSAP is more than a convenient array of symbols conveying uncertainties in technical information. It is a coherent 'notational system' which provides a general framework so that an unlimited variety of particular notations may be employed unambiguously. In form, NUSAP is a string of five positions corresponding to the categories of numeral, unit, spread, assessment and pedigree. By means of this place-value representation, each category can be expressed simply, without need for its explicit identification (this is a 'system' of notations at the most abstract level). For each category, there are many possible sets available for conveying particular desired meanings. Thus in unit we may have Imperial, CGS (centimeter-gram-second), MKS (meter-kilogram-second) or SI (Système Internationale) units, for different systems within physics. Any particular array of such sets we call a 'notation'. Given such a notation, any particular case of representation will be an 'instance' of the notation. Thus '5.21 cm' is an instance of a partial NUSAP expression in decimal arithmetic and CGS.

Such formal distinctions impart great flexibility and power in the expression of quantitative information. In this respect NUSAP is analogous to the notation system of 'Arabic' numerals, where the meaning of a digit depends on its place, so that a set of ten digits is adequate for the representation of any possible integer. By means of this flexibility we have a means of escape from the 'digital trap' of arithmetical representations, whereby even those notations used to qualify an expression are themselves afflicted by pseudo-precision. Our traditional notation makes this very difficult to avoid; thus if we write '95 per cent confidence limit', this does not ordinarily convey information significantly different from that expressed by, say '96 per cent confidence limit'; and yet when we write the expression digitally we are implicitly making such spurious distinctions unless we adopt rather special and cumbersome notations, as (in this case)

[9½]*10%. Comparing graphical representations with digital, we see how by their means pseudo-precision is almost always avoided, simply because the grid is not drawn over the whole area defined by the axes. The eye estimates a value, rather than being given precise (possibly pseudo-precise) information on it. With NUSAP we do not simply escape from pseudo-precision, but (with its aid) we can recognise and then manage it.

The NUSAP categories

Turning now to the separate categories, we distinguish first between the *numeral* and the *unit*; by numeral, we mean an 'arithmetic': a set of elements and relations which corresponds to analogous operations in the real system under consideration. The simplest case will be that of the traditional arithmetic of the counting of small sets and of the measuring of straightforward magnitudes. It might be thought that this arithmetic is the one which is commonly used in calculation, ordinary and technical, but this impression is incorrect. In order to control the inherent uncertainties even in ordinary quantities, various 'artefactual' arithmetics are in use. The most common example is the calculation with 'significant digits'. Thus in a multiplication, the 'extra' digits in the product are simply discarded, even though they might be the correct result, because the information contained in them is too uncertain to be useful or appropriate to retain. In the compilation and aggregation of statistical data, large numbers are allowed to 'swamp' small ones in a sum, when the small number is less than the range of uncertainty of the large one. Thus an archaeological object that was dated as 8,000 years old, some 12 years ago, is not now dated as 8,012 years old. In every such case, the artefactual arithmetic is introduced for pragmatic reasons and is governed by craft skills and tacit judgements, usually ignored by teachers and philosophers of mathematics.

We said before that the contents of the numeral place need not be ordinary numbers. Thus, if a quantity is known only to within an 'order of magnitude', then an appropriate instance of numeral would be *E6*:. We remark that an instance *1*:*E6* denotes a determinate quantity, a million, very different from the 'order of a million' conveyed by *E6*:. (Representations in NUSAP have the places in the string separated by a colon; in reading them, we express the colon by 'on'. For example, an instance *1*:*E6* — where *E6* is in the units box — reads '*1* on *E6*'.)

If one quantity is known only as an interval which lacks any preferred point of likelihood or of symmetry, then this should be the entry in the numeral place. Thus we could have (a,b): for an ordinary interval, $(\geq a)$: for an open-ended one. In the numeral place we may also find expressions of yet more general mathematical structures, as numbers of a finite set representing an ordinal scale (as in much of social research), or numbers representing indices or 'scores' with a purely artefactual arithmetic. An extreme example of an ordinal scale with a qualitative notation for numeral which is of direct practical use is that for Geiger counter readings, as 'click', 'chatter' and 'buzz', representing the sounds made by the counter, for low, medium and high rates of recording of radioactivity. When the meter goes up to 'buzz', that is a totally qualitative (and perhaps misleading) but none the less very emphatic message of urgency.

Unit is the base in which the numeral is appropriately expressed. We distinguish two components of the unit. There is the 'standard', the common or generally used unit of the relevant operations; and the 'multiplier', relating the standard to the particular unit involved in the expression. Thus we frequently see '£342M', where the unit (£M' (with '£' as standard and 'M' as multiplier) is the actual basis of the calculations reported, as distinct from the '£.p' of strict accountancy practice. The meaning of the pair standard–multiplier may of course vary with context; thus 'kg' is now a fundamental unit in the SI system, in which strictly speaking, '1g' should be written as '1mKg'. These two quantifying categories enable a refined description of topologies and scales of measurement.

We can illustrate the application of NUSAP on some simple examples, where existing representations are inadequate. Suppose that we start with 'five million', and we add some smaller quantity. If it is very small, as, say, 180, then the sum is normally understood still to be five million, since the latter quantity is not significant in the context. Writing the sum formally, we have 5,000,000 + 180 = 5,000,000. In this sum, the last three zeros are interpreted as fillers rather than true digits; and so we use an artefactual arithmetic, adopting implicit conventions for the neglect of certain digits, just as in rounded-off calculations. But if the second addendum is 180,000 it is not clear from the uninterpreted sum 5,000,000 + 180,000 just where the counter digits end and the filler digits begin. Only from the context can we know whether to apply a natural or artefactual arithmetic. A notation like 5×10^6 may help, but even that is not conclusive.

This distinction between 'counter' and 'filler' zero-digits is not commonly recognised. It is one of the fertile ambiguities in the 'zero' concept which makes our numeral system so powerful. One might even say that filler zero-digits are necessary for the system, otherwise a higher-order unit (for example, 'thousand') would have no meaningful expression in it, and the place-value would be inconceivable. Yet a sharp and clear distinction between fillers and counters for zero-digits would run counter to the ideology of the 'simple' and 'elementary' character of our number system (already quite implausible on historical grounds).

In NUSAP we can express 'five million' in various alternative forms, $5{:}10^6$, 5:M or 5:E6. Here it is explicit that the unit is millions. Although some ambiguity remains, it can be resolved by the entry in the spread position. But it is quite clear that 5:M+180 = 5:M is the correct sum, unless there is an explicit note to the contrary in the spread position. There is no need for an artefactual arithmetic with all its ambiguities.

NUSAP can also convey some shades of meaning that may be important in particular contexts. Thus 'five million' may be better represented as 50:E5 or ½:E7, denoting different sorts of operations in the different aggregated units. We note that the use of fractions in the numeral position enables us to express the meaning of a rough cutting of an aggregated unit;

thus a 'third of a million' is represented better as ⅓:M rather than $.33 \times 10^6$. It can be considered an advantage of a notation, that a user can represent, and even calculate with, an instance which expresses a perfectly clear statement of a quantity that previously needed a verbal form.

Another useful example from ordinary practice is counting in dozens. This shows more clearly the influence of the process of production of the datum, since in this mixed base there is no ambiguity between counter and filler zero-digits. Thus for 'eggs' we may have as alternative representations 4½:doz-eggs, 9:half-doz-eggs and 54:eggs. The first corresponds to the price, the second to the packaging, and the third to the use (in small quantities) in cooking. This example illustrates how an expression that may be appropriate in one context (integer counts for use) may be pseudo-precise in another (prices or packaging).

Proceeding now to the spread category, we again use aggregated counting, with the above-mentioned ambiguous case of 180,000 added to 'five million'. It may be that the larger quantity here has such inexactness that even a tenth of it is insignificant. This could happen if it is part of a sum with much larger quantities, as 32:E6: and 155:E6:. Then the spread would be understood to be as large as E6, the unit, and therefore the 180,000 or .18:E6 would be meaningless. In this way, the notation represents the practical situation of the swamping of a much smaller quantity in a sum. To be completely explicit we may express this as follows: 5:E6:E6. The spread, E6, indicates that no interpolation within the scaling has been done. Equivalently, every quantity in this sum has an inexactness interval which is E6 in length.

Strongly inexact quantities are sometimes expressed as 'within a logarithmic interval of 10'. We prefer that expression to 'to within a factor of 10' for this is ambiguous, possibly referring to a factor 10 on both sides of the number, leading to a logarithmic interval of 100, or perhaps referring to the logarithmic interval 10 itself. This expression fits the legendary 'first law of astrophysics', namely $1 = 10$. It would be pedantic and pseudo-precise to express the end-points of the logarithmic interval as $(10)^{1/2}$ more or less than the given quantity; a rough approximation, giving a convenient digit, is appropriate. Thus 3::Λ10 would be represented by the inequality $1 < x < 10$, while 5::Λ10 would be $1\frac{1}{2} < x < 15$. We can also express inexactness given in proportional terms: for example, '5×10^6' with a proportional error of 15 per cent is represented as 5:E6:15% or as 5:E6:[15 in 100].

The assessment category expresses the reliability of the information, generalizing not only the confidence limits of classic statistics, but also those of Bayesian statistics, interpreted as 'degree of belief' (Keynes 1921) or 'betting odds' (Savage 1954). Such formally defined measures are properly applicable only in special cases, and are not free of conceptual problems of their own. The assessment category is not to be formalised in the logical sense, but is designed to convey judgements of reliability in a convenient form. Where statistical notations are familiar and appropriate, they may be freely used. Otherwise, a more qualitative notation such as for example an ordinal scale should be adopted. Thus, we may have the set,

(Total, High, Medium, Low, None), perhaps codified as (4,3,2,1,0) to convey this kind of judgement. As in the case of numeral, unit and spread, a great variety of notations are available for assessment.

In traditional statistical practice, the assessment (or confidence limit) is closely associated with the spread (or variance). Unfortunately, this association tends to conceal the radical difference between the two categories, and to inhibit the understanding of either. When we generalise assessment from the simplest notion of reliability, the independence of the two categories becomes apparent. For example, consider a statistical distribution where we are interested in estimating the 95th upper percentile, or the top 5 per cent. The entry in the numeral place is then qualified by the expression '%95' in the assessment place, the order being inverted deliberately to distinguish this from the more traditional '95%' confidence limit. In such a case, the spread will depend on the number of trials or of simulations of the same process. So, if we are comparing the results of two different experiments involving different numbers of trials or simulations (as for instance obtaining the top 5 per cent of a distribution of experimental coin-toss results), we can have spreads varying with the size of the sample while the assessment entry is always '%95' (Funtowicz and Ravetz, 1986).

The categories of the NUSAP system are already in practical use in cases where the reputation of a providor of technical information depends on their recognition of the various sorts of uncertainty affecting it. For example, public-opinion polls are now typically given in the form of a percentage (for example, 42 per cent) agreement with a stated position; then an 'error range' (for example, ±3 per cent); then a confidence-limit on the range (for example, 95 per cent), and finally a note on the sample (this may include the number questioned, information on their location, age and social class, etc, and perhaps on sampling technique). Such a compound statement corresponds very neatly to the NUSAP categories; We can express it as:

42 : % yes : ±3 : 95% : Pedigree (Sample)

Such a sophisticated form of expression was developed in response to criticisms of polls for giving inaccurate or misleading results. It shows how the NUSAP approach applies naturally whenever quantitative information must be provided in a responsible manner for its effective use in policy-making.

Statistical information

Our first exercises in applying the NUSAP system were mainly in what might be called 'research' areas, where quantitative information is derived from a small set of experiences, either physical experiments or personal estimates. We now apply the system to another kind of technical information: the 'statistical' in the original sense of having to do with statecraft. The production of survey data is as old as organised government itself. In modern times, the power of analysed information was first demonstrated by

William Petty's *Political Arithmetic*, on the seventeenth century. Since then, census-taking has become a responsibility of all governments. In the case of the United States of America, it is related to the constitutional duty of the President to give an annual report to the people on 'The State of the Nation'.

Statistics have a contradictory reputation with the public. Although they are necessarily relied on, they are also distrusted as in the saying, 'lies, damned lies and statistics'. There is a growing technical literature that teaches 'straight and crooked thinking' with statistics (Jaffe and Spier 1987). The NUSAP system can be seen as a contribution to this development of quality-control in statistics.

For the proper understanding of statistical information, and its most effective utilization, we must appreciate its important differences from the 'research' case. Here, data are not collected for the investigation of some particular research problem, but in order to become part of a store of information available for possible future uses in a policy context. Their significance may be less in their absolute quantities than as elements for comparison in a time-series. Normally they are collected in widespread field operations, and are then subjected to a sequence of operations of aggregation and interpretation. The 'theoretical' component constitutes the framework in which the field-data itself are categorised, rather than furnishing hypotheses to be tested as in the 'research' case. All these properties make it necessary for the work to be done institutionally. We note that more than one institution can be involved in the process; and independent individuals can 'borrow' data provided by the original institution for review and publication (i.e. 'secondary analysis'). Such 'external' agents can be seen, for the purposes of the pedigree analysis, as simply another element of the sequence of operations whereby field-data is finally converted into statistical information.

When we consider the production of each piece of information in organisations, several features are relevant:

The context is of some 'mission' that defines the organisation, gives it a reason for existence and shapes its work and culture;

The work is done in such a way as to ensure consistency and continuity within the institution, and effectiveness for its outputs;

There is an explicit division of labour on all processes, and explicit definitions of individual tasks;

Every task is tightly embedded in structures of instruction and of evaluation;

The methods of work and even the concepts being manipulated are defined explicitly through superiors in the organisational structure;

The task of evaluation, accomplished informally in the 'research' case by the colleague community, is done formally here through review procedures, either internal or external;

A complementary set of unofficial networks and practices exists, which may modify or even nullify those of the official part;

There is no limit to the degree to which an institution can temporarily drift towards isolation from its external environment, be it the public to which it is nominally answerable or the ostensible object of its officially assigned tasks.

Of course, none of these features are unique to formal organisations. Even in traditional laboratory science, there is something of an hierarchical division of labour, and the 'bootlegging' of research and production of vacuous results are not unknown. But traditional science has operated in a style involving remarkably few formal structures; and philosophers and sociologists of science have generally neglected bureaucracy in the production of knowledge. Also, the really interesting science, traditionally, is that which has been successful. The phenomenon of an empirical scientific field having little contact with reality has not generally been considered worth analysing by historians. Hence the specifically institutional phenomena in the field of knowledge-production are still to be explored. In the present period, we may soon need to consider two intermediate cases: large, highly institutionalised research enterprises, notably in the defence area; and traditional sciences that are called on to supply results of a quality beyond their strength to produce, as in the assessment of environmental problems like low-level toxicants or potential future hazards.

For the moment, we shall concentrate on organisational structures at the opposite pole from traditional science. These are involved in the sorts of statistical information with which we are concerned. Complaints and uncomplimentary jokes about bureaucracies and their workings are doubtless as old as the institution itself, but we cannot assess the problems of this mode of operation unless we first appreciate its necessity. For this, let us take a very simple example of the definition of an object studied in the public-use context. Suppose that someone wants to know the 'total seating capacity' of the auditoriums in a city, perhaps for inviting a large international congress. A survey is done, but in the absence of uniform definitions, many kinds of things are included. Some enumerators simply count the chairs in each hall; others include standing-room and estimate the sitting-space on steps; others add the floor-space on stages where particular audiences (children, or Orientals) may be expected to sit or squat; some accept the local Fire Department's posted limits, others do not; some even argue about the definition of 'auditorium', on a continuous scale from an opera-house to a Scout-hut.

If each surveyor simply counts what is obvious to her/his common sense, an incoherent and useless number will emerge from the aggregation. To achieve an aggregation of numbers representing roughly the same thing, there must be prior decisions on definitions (related to the possible anticipated uses of the statistics), routines for ensuring uniformity among those in the field, and quality-checks of various sorts at every phase. If any aspect of this work of constituting and reviewing the inquiry becomes sensitive or contentious to any participant or observer, then standardised procedures for debate and decision must be available for use. There will also be the inevitable set of informal structures and practices, which must be appreciated by those whose use of the information is sensitive to its particular concealed biases and distortions. A reader might do a flow-chart for the production of some apparently simple sort of information, as in the example above; then the complexity and vulnerability of the processes of producing statistical information would be better appreciated.

Simply to have the same labels for data is no guarantee that the data will all refer to the same things. This is most clearly seen when data collections are separated by space, time or institutional location. When the indicators being measured are of a more abstract character (as 'seating' compared to 'seats') pitfalls in interpretation are more likely. Because of this, both comparison and aggregation of data sets require an elucidation of the operational meaning of their defining terms. In the above example, we could imagine either several cities adding their 'auditorium seating' data in spite of divergent standards; or a growth-rate calculated for one city over some years. Such variations may reflect differences in what is a 'reasonable' or 'appropriate' measure, in terms of existing practices or policy functions. For example, 'seating' will depend very strongly on the state of local concern with safety. To aggregate or compare data gathered under different circumstances of safety-consciousness could yield very misleading results!

Of course no set of established definitions and procedures can achieve complete uniformity among all collections of the 'same' data. Here we are not criticising existing practice against the standard of an impossible ideal, but establishing the case for the management of the inevitable uncertainties in statistical information.

Pedigree

We recall that the pedigree category in NUSAP is not only an evaluative account; it also has the function of expressing ignorance. Can we say anything useful about that of which we are ignorant? It would seem by the very definition of ignorance that we cannot, but the boundless sea of ignorance has shores where we can stand and do a map. Let us think of a 'border with ignorance' as the last sort of uncertainty we can now control in practical technical work.

To map this border in the pedigree, we describe the state of the art of the field in which our quantity is produced. This is done by a matrix displaying the degrees of strength of crucial components of the process. We distinguish between the different sorts of theoretical frameworks and of data-inputs. These range (in the former) from well-established theories to unsupported definitions, and (in the latter) from well-controlled experimental data down to uneducated guesses. Because we cannot always assume that the social dimension of the practice includes a fully matured field, our analysis of that is divided into two aspects: the peer-acceptance of the result and the state of paradigm development, or colleague-consensus, in the field.

For comparison we include the pedigree-matrix for research information (see Table 7.1). In that case there is an obvious reference to an epistemological taxonomy in the pedigree; the names of the various modes

Table 7.1 Research information pedigree-matrix

Score	*Theoretical structures*	*Data-input*	*Peer-acceptance*	*Colleague consensus*
4	Established theory	Experimental data	Total	All but eccentrics
3	Theoretically based models	Historical field data	High	All but rebels
2	Computational model	Calculated data	Medium	Competing schools
1	Statistical processing	Educated guesses	Low	Embryonic field
0	Definitions	Uneducated guesses	None	No opinion

Table 7.2 Statistical information pedigree-matrix

Definitions and standards	*Data-collection and analysis*	*Institutional culture*	*Review*
Negotiation	Task-force	Dialogue	External
Science	Direct survey	Accommodation	Independent
Convenience	Indirect estimate	Obedience	Regular
Symbolism	Educated guess	Evasion	Occasional
Inertia	Fiat	No-contact	None
Unknown	Unknown	Unknown	Unknown

and the scalings relate to traditional understandings of the strength of scientific information. In the case of statistical information, the philosophical problems are as yet little explored by scholars; hence the epistemological aspect of this pedigree matrix may be less obvious. When we consider how institutions accomplish the social construction of knowledge, we see that the complexity of the border between knowledge and ignorance requires well-articulated analytical tools. By their means we can describe how an institution can either achieve usable knowledge or create official ignorance, depending on its operations and internal culture.

Every entry in a table of statistical information has a complex history of its own; it is the product of both formal, reproducible operations and of personal judgements at various points. To provide a full historical account of each such entry, as an explanation of the table for potential users, would be an impossible task. The brief fine-print annotations on such statistical tables must normally suffice. The pedigree given here is designed to provide an evaluative account of the relevant aspects of the statistics, in a convenient codified form. It cannot convey the full account, nor is it designed for the explicit communication of judgements of craftsmanship and of the quality of the technical work. Rather, it expresses the salient modes of operation in the different phases of the production of information. The phases are: *Definitions* and *Standards, Data-Collection* and *Analysis, Institutional Culture* and *Review*. The pedigree matrix for statistical information is displayed in Table 7.2.

By the first phase of this pedigree matrix, Definitions and Standards, we understand: all those decisions, logically prior to actual enumeration and testing in the field, concerning the establishment of the relevant conceptual objects and the set of operating procedures. We have seen from the previous example of 'seating' that if data-collection is guided merely by common sense, then a totally incoherent set of numbers will emerge. This is of course a matter of degree: 'chairs fixed in a position' is a fairly straightforward object for counting; 'seats' is rather more abstract and therefore elastic; and 'seating space' is even more so. As we become more abstract in our concepts of what is to be counted, we need more clarity in what is to be considered as a candidate for counting.

Some objects of data-collection are of a compound character; their constitutive parts having all possible positions in the scale from the concrete to the abstract and the theoretically constructed. For example, in developing countries, major environmental threats come from bad drinking water. This is the object of the United Nations 'Water Decade' of the 1980s (Agarwal *et al* 1981), devoted to the provision of safe drinking water and adequate sanitation for all the population of the developing nations. Progress in this programme is officially measured by the indicator: 'proportion of the population (rural or urban) with access to safe drinking water' (WHO, 1984). Ordering these components from the simpler to the more complex, we have: 'water', 'proportion', 'population', 'rural', 'drinking', 'safe', 'access'. The first item is a natural object; the second a purely mathematical entity. 'Population can in principle be determined by well-known demographic techniques; while 'rural' may require some conventions (as on maximum size of a settlement for it not to be counted as 'urban'). 'Drinking', of water, relates to use of particular supplies, and has its effective meaning determined by the last two components. 'Safe' is in some respects a scientific category, but this judgement depends on a very complex process. It starts with scientific principles of bacteriology and toxicology, extends through standards as defined for laboratory practice, over to testing under field conditions, and also to quality-assurance on results of tests, and finally, includes judgements of 'acceptable risk'. In this case, the technical aspects of the term are intimately involved with institutional and cultural aspects. The final component of the indicator, 'access', is as abstract as 'safe', but lacks even its partial basis in scientific research. 'Access' relates so strongly to habits and expectations, and even to the cultural meaning of the activities of obtaining water, that any simple conventional standard (as, 100 meters from the household) is almost useless. Our rule for evaluating such compound indicators is generally by the weakest significant link. Thus, even if 'rural population' is quite well defined in some place, and 'safe drinking water' has some reasonable scientific meaning, 'access', a component to which any aggregate statistics are highly sensitive, will, for us, determine its characterisation for pedigree.

Under the heading Definitions and Standards we have five modes: *negotiation, science, convenience, symbolism* and *inertia*. The mode 'science' refers to those cases where the definitions and standards are firmly based on some background body of scientific knowledge. This occurs rather less often than might be thought. Other considerations, particularly local interests in a policy problem as well as the limits of practical feasibility, can render the strictly scientific approach inappropriate. Thus there is a use for negotiation among interested parties, where the existing scientific knowledge needs to be adapted to local circumstances. Where the background science is inadequate for a full specification of the definitions and standards, several alternative modes are possible. We speak of convenience when those actually doing the work adapt any externally imposed definitions and standards to their particular circumstances. Sometimes, however, the scientific basis is nearly, or completely, irrelevant to the interests of those who ultimately control the institutional task. The need for institutional legitimacy or prestige, or other manifestations of values, may come to dominate over all other considerations; we speak then of 'symbolism' as the mode in this phase. It can be detected by the presence of definitions and standards, which, however technical their form, are known to be wildly inappropriate or simply irrelevant to the task. There is a further mode, 'inertia', to characterise the state where no one now even knows or cares why the particular definitions and standards came to be used.

The order in which the modes are displayed is roughly normative; clearly science and negotiation are better than inertia. We put negotiation over science because it represents a richer contact with the real policy dimensions of the task. Symbolism is fairly far down, because it can produce results with very little real content. However, these rankings are rough, and the proper evaluation of the information is built up through all four of the loosely related phases of this pedigree.

The second phase, Data Collection and Analysis, refers to the operation in which field data is gathered and then analysed and reported. Such work is inevitably a sequential operation; analysis comes after collection and will itself usually include several distinct levels. There are three sorts of possible practice in the measurement of policy indicators through field data. It may be impossible or unfeasible to measure directly (as, for example, total annual mileage of cars in some country); hence there can be an *indirect estimate* using some other variable related to the indicator by various intermediate assumptions (in the example, total fuel consumption and a standardised mileage conversion factor) (Mosteller 1977). Of course it is sometimes possible to have a *direct survey*, as in a census, where various policy indicators are measured all at the same time. Every inquiry benefits from being completely designed around its own particular problem, with special care taken for characteristic pitfalls. Thus, paradoxically, a *task-force* dedicated to an intensive partial study of a problem, may yield better information about the phenomenon in all its complexity and variability, than a route census survey which is nominally comprehensive. Coming to the other end of the scale it may well be necessary to provide some information even when field data is simply unavailable; for that, an *educated guess*, based on expert opinions, may be appropriate. Finally, we have *fiat*, where some authority decrees the number which shall officially represent the phenomenon in question.

The subsequent phases of this pedigree refer explicitly to the organisational character of the exercise. Institutional Culture characteristics relations between different elements in the sequential operation. In most institutions there is a formally defined hierarchy, where instructions go 'down' and information goes 'up'. In the present case, definitions and standards come from 'above' and field data from 'below'. Superficially, upper levels are related to lower through instruction and review; and the inferiors are expected to have obedience to their superiors. This is rarely seen in the pure case, and there are two contrasting sorts of modifications. Sometimes there is a two-way flow, *accommodation*, resulting in a *modus vivendi* whereby tasks are performed more or less to everyone's satisfaction. Where this is not possible, inferiors will resort to *evasion*, where only the formal aspects of the tasks are satisfied, to the detriment of their real content. The extreme case of this phenomenon is where orders from nominal superiors are simply ignored; we call this *no-contact*. There is another extreme, where the formal hierarchy is strongly modified and softened by shared interests and open communication; this we call *dialogue*. We give the best rating to this, analogous to negotiation in the first phase.

In the last phase of our pedigree matrix, we analyse Review. In traditional scientific practice, this is accomplished through largely informal procedures, as, the peer-review of projects and the refereeing of papers. In market-sector enterprises, 'control' is partly on quality of product, exercised through special units (and also by purchasers), and partly on process, through 'audit' functions. Quality-control is increasingly recognised as important in all productive activity, thanks to the challenge of Japanese industry. They have also shown that high quality of work cannot be attained through simple obedience, but needs commitment and dialogue among all levels of an organisation.

Every properly run organisation will include arrangements for a regular review of its work. Normally this will be done by some special branch, operating routinely. For certain purposes, an independent review is used; typically this is done when unbiased judgements are important for the assessment of some product or process. In special cases, the review might not be merely staffed by outsiders; it might be organised on an external basis. This would occur when the organisation as a whole, or some particularly crucial or controversial aspect of its work, is subject to question. Coming down the scale, we have a situation where review, or audit, is at best occasional or indeed where there is none. In the absence of review, there is ignorance about the quality of the products of the task;

the evaluation of this phase should then make no positive contribution to the assessment rating of reliability.

This phase, Review, rounds off our picture of organisational operations as conveyed by this pedigree. It is complementary to Institutional Culture, as both relate to the institutional style; while Definitions and Standards and Data Collection and Analysis relate to the work. It is particularly important to include this last phase in a study of bureaucracies, for of all organisational cultures they are the most prone to drifting out of contact with reality. An institution that lacks means of enforcing review and revision of its tasks and methods will inevitably become isolated and self-serving, and will eventually lose its legitimacy.

Practical problems

For the sake of simplicity we have assumed up to now that the data have been collected in the field, and then sequentially processed up through different levels of a single organisation. In many important cases, users of information are external to the institutions where the data are collected. The pedigree matrix developed here also applies to that situation. External users of information provided by an organisation can have a great variety of relationships to it. If they are in a superior institution, as a central government agency with subordinate though distinct regional agencies, then the original interpretation of phases and modes applies. The hierarchy may be weaker than this, as in the case of federal governments (such as the American, German or Canadian). It may be a strong transnational organisation like the EEC, a weaker one like the United Nations, or an association like OECD. In these examples the relationships of authority are progressively attenuated. They are quite absent in the case of individual or corporate users of information, operating outside the institutional nexus. Such persons might be experts solving particular problems in a symbiotic professional relationship with the institution, or they might simply be readers of published works where the information is contained. These last elements in the sequence of processing can handle the data in many ways. If they merely copy or report them passively, there is no change in the pedigree. On the other hand, they may manipulate them or even interact with providers at any previous stage of the sequence. These various possibilities should be reflected appropriately in the pedigree of the information they produce.

The problems of aggregation of incoherent data-sets are reflected in the pedigree matrix at several points. It frequently happens in practice that those who aggregate data find themselves with data-sets which are incomplete or incoherent in some way or other, and where they have no power to improve this material. They are inevitably forced to make policy decisions about the 'best' way to render such imperfect materials in the aggregation process. Perhaps the simplest problem is when one constituent is so large compared to the others that its contribution, and eventually its changes, make all the others irrelevant. An example is India in the context of its neighbours in South Asia (Appleton 1984). Whether one aggregates large and small constituents, depends on the context; if all that counts is the total, regardless of the distribution, then aggregation is justified. But users of the data should be made aware that the smaller constituents are swamped by the larger, and hence no effective information is being conveyed about them. Uncertainty concerning their statistics is drastically increased.

More complex problems arise when data-sets have inconsistent definitions or are incomplete in some way. Then aggregation and comparison becomes a skilled craft; those who do it must be quite sensitive to the costs and benefits of any particular procedure for possible classes of users. It is in this problem that we can see the close relationship between the different phases of the pedigree for statistical information. Any imperfections in Definitions and Standards and lack of communication reflected in Institutional Culture, must be managed in Data Collection and Analysis; and finally scrutinised in the phase of Review.

The various modes appearing in the four-place pedigree category may be most simply represented by a numerical code, from 0 to 4 or '?'. These codes are not simply normative. Although there is a rough normative scaling implied (as the explanations above have shown), we can see that no single mode reliably represents the whole history. Taken together they provide an analytic tool (as other empirical indices do) for the general 'style' of the production of the information. This overall characterisation clearly relates to the reliability of the information in use; hence we can express this as an index in the assessment category of the NUSAP system. A useful ranking system could be:

Total score (out of 16)	*Assessment code (reliability)*
≥ 13	High
≥ 10	Good
≥ 7	Medium
≥ 4	Low
≥ 3	Poor

The assessment and pedigree categories, with convenient representations, are displayed in Table 7.3.

An illustrative example

We can illustrate the uses of this pedigree by a simple example: the introduction of hand-pumps for drinking water in rural areas in a less-developed Third World country (Pacey, 1977). In this example, elements of several histories are combined, to provide a plausible reconstruction.

Our story starts long ago, when a ministry in a newly independent state was requested by some United Nations agency to report the number of pumps in rural areas. Lacking any means to study the phenomenon on the ground, the civil servants eventually located some numbers in files left

Table 7.3 Assessment and pedigree categories

$\Sigma \geq$	*Assessment* *Rel.index*	*Code*	*Def.&St.*	*Pedigree* *D.C&A*	*Inst. cult.*	*Review*
13	High	4	Neg	Task	Dial	Ext
10	Good	3	Sci	Dir	Accom	Indp
7	Medium	2	Conv	Indir	Obed	Reg
4	Low	1	Symb	Guess	Evas	Occ
0	Poor	0	Inert	Fiat	No-c	None

behind by a previous regime. These were duly conveyed as 6,970. Were we advising a user, concerning the reliability and history of this number, we would first need to elicit the information enabling us to identify the appropriate modes. Supposing this done, we could express the quantity as:

6,970:pumps: :Poor:(Inert,Indir,No-c,None).

We notice that even in such a case an adviser can transmit an official statistic while expressing due warning on its quality.

Later a regional association organised a conference on 'rural developments'. All participants were to supply figures on pumps, among others. The new government of the country was not to be worsted by his neighbours, and so a report, complete with precise numbers, was drafted. Here the full NUSAP expression could be

15,432:pumps: :Poor:(Symb,Fiat,No-c,None)
or 15,432:pumps: :1:(1,0,0,0).

As the country developed in strength and institutional maturity, a national census of village pumps was organised. Enumerators visited very village, complete with instruction manuals and clip-boards. However, there had been an inadequate training and organisation for the exercise; erroneous reporting and aggregation was common. We can express the result as

11,767:pumps: :Low:(Conv,Dir,Evas,None)
or 11,767:pumps: :6:(2,3,1,0).

(We note that 'Reliability Index' here would be 6 or Poor; still an improvement.)

A few years later, a similar census was organised, but designed to avoid the mistakes of the previous one. Some experts were brought in to advise on the organisation, and the Army was used for the field work. The aggregated number reported in this case was 7,566±46; or in the NUSAP system

7,566:pumps:±46:Good:(Sci,Dir,Obed,Reg)
or 7,566:pumps:±46:10:(3,3,2,2).

We notice that this is the first case in which a spread entry is appropriate.

Knowing the sensitivity of an impending policy decision to the relevant properties of the number, we can adjust the NUSAP expression accordingly, to avoid hyper-precision. Turning to the qualifying end of the expression, we remark first that the Good strength in the assessment category entitles us to take the reported error term as the basis for the spread category. In this last case, the most rigorous expression for N, U and S would be 75:[100's pumps]:½×100: , but for the sake of easy comprehension, this might be written for publication as 7,500:pumps:50: . For the policy-maker, it could be sufficient to report 7,500:pumps:50: Good: . When the reliability is good, the precise pedigree is not always required; when it is poor or nonexistent, then a scrutiny of the pedigree may be the only way to make the information useful.

Eventually, the country joined those benefiting from international aid for development. Hand-pumps were the subject of a special bilateral programme, largely administered through the private firms providing the equipment. At its formal conclusion, there was not much new to be seen on the ground. How to report that successful outcome which was so strongly desired? The technique was simple: all the items classed as 'pump' on invoices and order forms were added to the results of the previous census. The supposed total provision of pumps was conveyed by a number whose full critical NUSAP expression would be

11,402:pumps: :Low:(Conv,Indir,No-c,None)
or 11,402:pumps: :4:(2,2,0,0).

Let us briefly review what was understood as 'pump' in each of the above cases, and consequently how the processing was done. In the very first case, 'pump' was no more than an entry in an old file; hence we considered the mode of Definition and Standards to be inertia, while the number itself was at best an indirect estimate. In the second case, however, the symbolism swamped all other considerations; that to which the number referred was not pumps but a persuasive fiction. The first census enumeration was conducted on a common-sense basis. A 'pump' was what each enumerator thought it to be, within a broad class of devices with a wide variety of degrees of operability. Hence we call that convenience. The second census, better organised, had some science in the definition of pump. Those which were either not installed, or if installed, abandoned, were not to be counted in the enumeration. Thus at last some element of disciplined realism was brought into the process. But the temptations of progress soon worked their way in, and in the concluding report on the development contract, 'pump' once again became a fictional unit of accounting. Since the entries on the invoices in this last case had some relation, however tenuous, to objects on the ground, we describe the mode as indirect estimate rather than fiat or guess.

To continue our narrative, we imagine that this last set of statistics for pumps (derived from invoices and order forms) was sent to the regional headquarters of an international funding agency. There they were quickly recognised as nonsense. After a brief consultation with ministry officials, some plausible substitutions were made. For this operation, the pedigree could be (Conv,Guess,Dial,None). Thus dialogue can occur even when all other phases are vacuous or nearly so. Next, this little helpful exercise was discovered by Head Office because of a 'leak'. Aware of their vulnerable position as international civil servants, they insisted that the numbers on the original submitted form be accepted, and so they were; the pedigree would then be (Symb,Indir,Obed,Ext). This presents us with an apparent paradox, for the 'Reliability Index' in this case would be 9 or Medium. However, this case serves to remind us that the reduction of qualitative judgements to quantitative statements cannot be an automatic procedure. The present case is (hopefully) exceptional. Any competent expert noticing the unusual array of modes would already be alert to the possibility that simple addition of codes would yield a misleading figure in assessment and would write Low for that category, so that users would be warned that something was amiss.

More recently, a proper bilateral aid agreement was negotiated for improvement of water and sanitation facilities in rural areas. Survey work was of the best international standard in its technical aspects. Field enumerators were well trained in reporting conventions, but these were not adapted to local social and cultural conditions within the country. Given their competence and commitment, they were able to make an informal accommodation between the international conceptual structures and local realities. Therefore, we can give a pedigree (Sci,Dir,Accom,Reg), in this case, or (3,3,3,2), with a Good assessment. When their report was received at the planning office of the bilateral aid agency, the experts were in something of a quandary. The programme was under close scrutiny, and the directors were concerned to avoid the suspicion of having understated the existing provision so that their programme could later claim success. Hence they wanted a number that would be safe from understatement. Supposing that the original reported quantity was 11,300:pumps: 5%:Good: , they could translate this to 10½:[1000's pumps]:≥:High: . The coarser topology in unit and the minimum expressed in spread together enabled the higher assessment rating to be made.

At the end of the programme, the agency was required to provide an independent evaluation of its results before it could obtain a renewal from its sponsors. A highly expert independent team was sent to review the situation in the field. Their report was disturbing: many of the pumps seemed to be in a state of disrepair or even disuse. Moreover, in discussions with locals they found that many of the designs were inappropriate for local conditions. In one case the pumps caused marital strife in villages when wives refused to operate them on the grounds of their being so difficult as to cause ill-health. Nor did general health improve, as in those conditions the pump's water was frequently as polluted as that from open sources. For their evaluated report, the definition of 'pump' became crucial. Should they refer to all installed pumps, or to all those deemed operative, or only to those acceptable to the users, or (still more restrictive) those also provided safe water? By the time such problems of definitions were comprehended, the survey was underway, too late to be re-structured. In the report, the technical definition of pump was still used, but the tables were accompanied by extensive prose discussions of the limitations and pitfalls of the quantities reported. Here our pedigree reads: (Sci,Task,Dial,Indp), with High Assessment.

In spite of these troubling findings, the programme as a whole was considered worthy of further support and improvement. The findings of the 'task-force' survey were to be built into the specifications for the next phase of the programme. This proved to be not at all simple in practice. A number of conventions had to be agreed as to when a technically imperfect pump was to be counted as 'operative'. More difficult were such working definitions as those of 'convenient' and 'safe'. As we have seen, some depend on sociological and anthropological data, others on debatable scientific information. To achieve a realistic set of definable categories, the organisers found it useful to involve the local people themselves in some of the discussions. Hence the ongoing monitoring programme for the new project required a lengthy and expensive design process of its own. The pedigree for the information derived from that is (Neg,Task,Dial,Reg), with High Assessment.

In this hypothetical story of the evolution of hand-pump information, we have used pedigree as an analytical instrument in our discourse. In many or most of the cases, the user of the information would find the pedigree code sufficient for her/his diagnosis of the information. In some of the cases, the codes assigned could well be disputed; a resolution would then be achieved by a process of elicitation from the (fictional) producers of the information.

Conclusion

As we introduced NUSAP it was presented as a notational system or scheme, designed for the more effective communication of the uncertainties of quantative information. Having gone through the full explication of NUSAP and its application to an illustrative example, we can achieve a broader perspective. The notational system is the core of what we call 'the NUSAP approach'. This involves first of all, the elicitation of uncertainties from those who have produced the information. In the course of this, the basic concepts and methods (for example, seating, surveys) are elucidated. The representation of the quantity can then take on a variety of forms, each designed to be most effective in a particular context of use. Out of all this, we achieve means for a higher degree of quality-control on technical information, relating both to its production process (recorded in pedigree) and the product.

Through the NUSAP approach we may be able to overcome, or at least reduce, traditional public scepticism about 'statistics'. This scepticism is now a serious component of policy-making and communication on environmental and industrial hazards. One important contributing cause to this loss of credibility is the difficulty of expressing uncertainties in expert information. In the absence of an appropriate language, the concepts cannot be communicated, and indeed cannot even be grasped without awkwardness and confusion. As a result, technical information is usually presented in an overly precise form, lacking the essential qualifications about its uncertainties. When the quantitative parts are eventually discovered to be systematically off-target, all the information, and those who produce it, lose credit and credibility.

The NUSAP approach does not merely provide a more sophisticated and nuanced means of expression for the effective communication of uncertainties. In addition it provides the language whereby the experts themselves can achieve clarity of expression and thought, and also protect themselves against the inevitable pressures to provide simplistic, overly-precise information. By its means, quality-control can be exercised within all the communities involved in the use of information for policy.

As the NUSAP approach is brought to bear on actual processes of policy-making, we discover that no real decision depends on the precise numerical value of technical inputs. The various sorts of uncertainty may be quite as important as the numbers or (in the case of high spread or low reliability) may legitimately be taken as the dominant part of the policy-input. Without NUSAP, all this may seem to a technically trained expert as the intrusion of politics or subjectivity over rationality or objectivity. With NUSAP, this is a natural extension of the good practice that is traditional even in matured experimental sciences.

The NUSAP approach is a formal representation of a concept of knowledge which we believe to be essential for framing the role of science in the face of environmental threats. This is that uncertainties of all sorts cannot be banished from the research or even the solution of scientific problems. Rather, the proper management of uncertainties is the key to the design of an effective science for this crucial and strongly contested field.

The NUSAP system and approach have already been applied to a variety of issues. In this chapter, we have concentrated on one important form of technical information, the statistical, which clearly exhibits the problems of complexity and vulnerability, and the need for quality-control. Applications to other substantial issues, including some of direct environmental concern, are now underway and will be reported in due course.

References

Agarwal, A. *et al.*, 1981, *Water, Sanitation, Health, Space-Space for all?*, IIED, London.

Appleton, B., 1984, *Behind the Statistics*, WHO, Geneva.

Funtowicz, S.O. and Ravetz, J.R., 1986, 'Policy-Related Research: A Notational Scheme for the Expression of Quantitative Technical Information', *Journal of the Operational Research Society*, **37**, 243–7.

Funtowicz, S.O. and Ravetz, J.R., 1987a, 'Qualified Quantities — Towards an

Arithmetic of Real Experience', in Forge, J. (ed.), *Measurement, Realism and Objectivity*, Reidel, Dordrecht, 59–88.

Funtowicz, S.O. and Ravetz, J.R., 1987b, 'The Arithmetic of Scientific Uncertainty', *Physics Bulletin*, **38**, 412–14.

Jaffe, A.J. and Spier, H.F., 1987, *Misused Statistics*, Dekker Inc., New York.

Macgill, S.M., 1987, *The Politics of Anxiety*, Pion, London, 137–71.

Macgill, S.M. and Funtowicz, S.O., 1988, 'The Pedigree of Radiation Estimates', *Journal of the Society for Radiological Protection*, in press.

Keynes, J.M., 1921, *A Treatise on Probability*, St. Martin's Press, 1952, New York.

Mosteller, F., 1977, 'Assessing Unknown Numbers: Order of Magnitude Estimation', in Fairley, W.B. and Mosteller, F. (eds), *Statistics and Public Policy*, Addison-Wesley, Reading, Mass.

Pacey, A. (ed.), 1977, *Water for the Thousand Millions*, Pergamon, Oxford.

Savage, L.J., 1954, *The Foundations of Statistics*, J. Wiley, New York.

World Health Organization (WHO), 1984, *The International Drinking Water Supply and Sanitation Decade: Review of National Baseline Data*, WHO, Geneva, 154.

Chapter 8
Building Public Concern into Risk Management

Brian Wynne

Introduction

Several chapters in this book expose an issue which has also arisen in other social research on risk perceptions. This is the problem that ideas of risk which people have and use are part and parcel of the full flux of social interaction — they are embedded in concrete but diverse social contexts, in the maintenance of coherent moral existences in the broadest sense.[1]

Engineering and economic approaches to risk management have attempted to abstract general 'laws' of public risk perception in order to derive practical decision rules for acceptable risk. Starr's attempt to establish the 10^3 multiplier for voluntary risk taking over involuntary risks is an example.[2] Such abstract, quasi-scientific and (in aim) universal definitions of risk have contrasted sharply with the findings of social research. This has intended to the opposite extreme, to conclude that 'risk' has meaning for people only within their own ethnocentric cultural settings.[3] In surveying this spectrum of approaches (with psychology ranging between) one is reminded of Valery's dictum[4] that 'everything that is complex is useless; everything that is simple is false'. Social research looks to be nearer to the truth, but also appears as useless indulgence. However the simple attractions of definitions of risk which ignore the social context can no longer even claim to be useable. Thus some attempt has to be made to open up decision methods and processes to the diverse social contexts from which risk definitions arise and in which they are used.

Decision techniques such as benefit-cost analysis (BCA) and multi-attribute utility analysis (MAUA) have attempted to respond to public concern in systematic, structured ways. In this chapter I address the question of whether expanding such decision techniques to include factors for 'concern' can be expected to produce risk management decisions which are publically accepted. Here I can only outline the approaches, and compare their definitions of the well-springs of public concern with those derived from empirical social science research. I have done this in greater detail elsewhere, so that this is a somewhat condensed argument.[5] The central question is whether elaborated formal decision techniques can be an adequate substitute for institutional changes in decision-making as a response to public concerns about management of risks.

Formal decision analysis

There are of course various branches of decision analysis, and this is not intended to be a complete or definitive review.[6] The field combines economics and psychology to generate normative frameworks for defining rational solutions to decision problems. Decisions are treated as discrete occasions with a 'given' decision problem. There is usually only a single 'decision-maker', and collective values are assumed to be linear aggregates of separate individual ones.

The basic method is to define a branching set of decision options and probable consequences, attaching subjective values to each of those expected outcomes, and estimated probabilities of their occurrence. In risk analysis, the potential consequences of a technology can be factorised into different dimensions: death (to workers or public); morbidity; environmental damage; economic damage (for example, air pollution on crops or clean-up costs); etc. These can be further differentiated: for example deaths to children can be said to involve more damage than deaths to old people, and collective deaths in one event may be deemed more 'costly' than the same total number spread over a larger number of routine events. These are the 'attributes' identified in risk perception work. By weighting their 'negative utility' and aggregating them, an overall 'index of risk' can be composed of these factors for the activity in question. The choice of factors and their weighting are laden with evaluative commitments.

When units are attached to quantitative risks, many value-choices are made, whether or not these are explicit. For exactly the same activities or 'risk', deaths-per-worker-employed may give a very different comparative risk picture from deaths-per-unit-of-outputs or of time, of resources, or of capital inputs. These value-problems are familiar enough and have been tackled by several authors. A leading approach is that of Watson, Fischhoff and Hope.[7] They point out, correctly, that choosing an index of risk for a decision analysis is a value-laden, political act. This can never be avoided, but it can be made more transparent, considered and democratic. Decision-makers rather than experts can choose which attributes to use, with what weighting, in a composite definition of 'risk'. The weighting distribution across the range of chosen attributes would be the utility function since it expresses the negative utility for that decision-maker of those aspects of harm represented by the chosen attributes. If delayed death is evaluated more seriously than immediate death or injury to children considered more serious than a few years' life-shortening, these can in principle be represented openly in the index of risk. Having performed this evaluative exercise in an explicit way, for different competing options, the decision-maker can then trade off benefits and risks to arrive at his or her optimal choice. In theory, given sufficient accountability and consensus, this could be a societal choice. Thus these methods have gained credence as risk management public policy tools.

MAUA and BCA differ significantly in that BCA aggregates different attributes (usually a narrower range than MAUA) into a single definition of 'detriment', which is used as a single technical factor. MAUA on the other hand keeps as many attributes as possible, disaggregated, to allow the decision-maker access to change value-weightings of these attributes. Both methods have been used to try to incorporate public concern, but in different ways. BCA attempts to enlarge its single expert-defined detriment function by introducing factors for public concern such as those derived from 'willingness to pay' surveys.[8] Once inserted, these are largely fixed. MAUA attempts to keep the evaluative role in the hands of decision-makers, exogenously choosing different attributes and weightings. These technical structural differences reflect the different institutional roles which each technique plays in practical decision-making.

BCA is used to optimise for example radiation shielding, or discharge-control authorisations and associated investments. Traditionally in this field the cost or detriment function has been defined by statistical increments of risk of death at prevailing dose levels combined with a cost of such a death. However it was recognised in the 1970s that experts were making social value judgement in this process.[9] Instead of inserting exogenously defined values of life (and morbidity, etc.) in the detriment function, the 'democratic' logic of BCA is to ask the persons at risk, 'what are you prepared to pay to avoid a (defined) increment of risk?' Bodies such as the Health and Safety Executive (HSE) and National Radiological Protection Board (NRPB), which have to give technical advice in such practical areas, have been developing 'willingness to pay to avoid' data from questionnaires.[10] At present however, optimisation such as that implementing the 'as low as reasonably achievable' principles of radiological protection employs detriment curves which notionally include a 'risk aversion' factor (the slope of cost increases as the individual dose limit is approached) but which do not claim any empirical relationship with the inherent structure of public risk perceptions.[11]

The significant thing about the use of such BCA methods is that they are normally used in decisions 'at the technical level by design engineers . . . busy and hard-headed people [who] need reasonably simple and straightforward tools to assist them'.[12] These are relatively routine, semi-private decisions, where it is assumed the fixed evaluations made by experts in their composite detriment function are not controversial. Indeed when BCA has been used in more complex and controversial areas (i.e. where public concern does exist), the less explicit 'other factors', which are always acknowledged to play a role alongside formal BCA methods, come to predominate, and confuse the apparent clarity of decision rules. The setting of marine discharge limits at Sellafield is a good example. Also, the use of BCA in the Roskill Inquiry into the Third London Airport, and in motorway planning, was heavily discredited because as soon as conflict led to demands to justify the logical and evidentiary details of a BCA framework, it simply fell to pieces.[13]

Thus BCA appears almost by definition to be a technique which cannot be asked to tackle decisions involving a significant element of public concern or 'active' risk perception. In other situations, and as a carefully limited part of broader methods such as MAUA (see below), it is a different matter. Because of the inherent difficulties even of 'stretched' BCA in more conflictual decisions such as airports, motorways, nuclear waste sites and Sellafield discharges, MAUA has been taken up more, and has indeed been used in developing the 1986 BPEO statement for choosing nuclear waste facilities.[14]

Public perceptions as objective attributes?

There is no reason in principle why the formal procedures of decision analysis should not include an item for 'public concern' or 'risk-aversion' as part of the risk-cost indices. Watson *et al* expressly recognise 'concern' in this way, and advocate its inclusion (assuming it is not based upon simple ignorance of risk magnitudes) as one attribute alongside others in a multi-attribute utility analysis.[15] If there is no mutual dependence of preferences, and 'risk equivalence' prevails (i.e. one death is equal to 0.1 chance of ten deaths) then the aggregate Risk R of a given option may be written simply as:

$$R = \sum^{n} w_i \, u_i$$

for n separate attributes, where u_i is the expected utility for attribute i, and w_i is a weighting factor for its relative importance. 'Concern', once converted into units, for example of money equivalence, may be weighted in alongside other attributes as mentioned earlier — potential deaths and injuries of various kinds, environmental damage, etc. These negative utilities are termed 'detriment' in risk-benefit optimisation techniques. Converting 'concern' into units has been increasingly performed by eliciting people's willingness to pay to avoid a given risk increment.

As far as it goes, the stretched (Watson *et al*) MAUA which includes 'concern' is an admirable attempt not to patronise public risk perceptions, and to take them seriously as a substantive, legitimate part of risk management and policy-making. There are, of course, familiar and not at all trivial questions to do with reducing complex values and experience, into money terms, and with aggregation of qualitatively different kinds of 'attribute' and experience into single scales. However I want to identify two further problems.

The first problem is that the whole MAUA framework of Watson *et al* takes it for granted that the technology or system which is the risk-source is an objective, bounded entity, definitely separable from others and the same for all social groups. As outlined in chapter 3 this is never so, even for experts. The interpenetrations of any 'given' risk-system into other technical and organisational systems are manifold, and its boundaries are intrinsically ambiguous; indeed they are socially defined, not only as a matter of subjective perception, but of objective social experience and relations of the technology. It is therefore not only a question of what kinds of detriment (including concern) to incorporate from an assumed given risk-source, but what are the boundaries and shape of the risk-generating problem in the first place? As I suggested previously, it is implicit differences on this question which lie at the heart of public conflict and lack of authority for expert risk assessments which commit themselves to a particular set of assumptions about the boundary of the issue.

Because their objective social experiences are different, different culturally determined perspectives will define not only what risk-attributes are involved, with what weighting or utility, but the 'technology' (and social relations) involved. The examples given in my previous chapter, especially perhaps the 2,4,5-T case, emphasise the important fact that 'objective' risk analyses were based upon subjectively defined risk-systems which arbitrarily excluded relevant dimensions of the problem.

Essentially the same point about fundamental problem-framing beneath 'objective' risk analysis is enlarged upon below. I do this by reinterpreting conventional psychologically derived views of 'dread' and 'unfamiliarity', which are regarded as objective attributes of public concern associated with the physical properties of technologies.[16] I see them instead as symptoms of different implicit social assumptions which frame definitions and thus assessments of risks. Translated into terms of social relations, the usual decision-analytic framework, which has a process in which 'technology' creates 'consequences' or 'expected consequences' which are then 'evaluated', is exposed as fundamentally inadequate. It suggests a detachment from the concrete experience of technologies which is not true either for the public or experts. This mistakenly allows the belief that the full range of different social experiences and evaluations of risk and technology can be encompassed by elaborating the factors of method, rather than changing the institutions using such methods.

Although it expressly allows for a certain range of value-choices, MAUA responds to public concern by incorporating it into a decision technique which still requires a single, discrete, context-free problem. This 'required' problem-definition imposes a fundamentally inflexible value-framework around the choices which are allowed. This appears to allow more value-choice, yet radically limits it because it cannot recognise the social contextual properties of technologies and risks, nor the basic origins of concern in social relations. The fundamental structure of the method drives out any such context. As noted in Chapter 3, risk analysis inadvertently 'freezes' a particular view of that context into the analytical structure. It helps to clarify this point by using examples, as I do later. However, it is also important to understand it as a caricature of scientific method so as to see the general implications of this attempt to bring 'rational' order to a disorderly domain. Scientific reasoning begins from unrealistic assumptions about the world — for example of friction-free surfaces, perfectly elastic solids or perfect harmonic oscillators. It has to make such simplified assumptions even to begin to construct ordered knowledge. But as the example from Chapter 3 show, this normal process in science leads to problems when the knowledge is applied in human contexts. To assume that risk perceptions are about, say, a single waste disposal plant and then to look for 'attitudes' to that risk from which concern is derived may miss the whole dimension of perceptions and evaluations in the context that 'risk' as defined, a context which the scientific approach has assumed to be natural and unproblematic. In that context concerns may be about the future growth of disposal volumes, or the quality of management and regulation, it is a distortion of the meaning of concerns such as these to reduce them to attributes within the original limited frame which assume, like scientific approaches generally, that the frame itself is natural.

Public concern out of context

A second problem with MAUA therefore is that — like the generally predominant approach — it cannot recognise the context-rooted nature of risk-problems. For decisions involving conflicting values it is clearly an advance over BCA, but the question is whether it can incorporate public concern enough to allay that concern in practice. I want to give a brief example to illustrate this limitation.

I will first describe the conventional psychometric way of dealing with the 'dread' perception of nuclear power which feeds into the MAUA framework. I will then offer a different interpretation of the dread factor which highlights conflicting inherent social assumptions as the root cause of the divergent 'technical' risk analyses of experts and lay publics on this attribute. If there is any truth in my interpretation, methods which suppress these social judgements and offer no public discourse within which they could be negotiated are inherently counterproductive. The elaborated decision methods we are discussing would not achieve this on their own, but would need associated institutional developments in order to do so.

As already indicated, psychological risk-perception studies have attempted to factorise the multiple elements of risks which people supposedly perceive, 'weight' and accordingly react to. One of the most sophisticated and influential research traditions in this field is that associated with Slovic, Fischhoff and collaborators.[19]

In a 1979 study, 30 hazards such as x-rays, nuclear power, vaccinations, skiing, contraception and so forth, were related on nine characteristics — voluntary, controllable, known to those exposed, familiar, known to science, dreaded, certain to be fatal, catastrophic and immediate in effects. Four different sample populations were used: the League of Women Voters, businessmen, students and a small group of experts knowledgeable in risk assessment. Their responses were factor analysed, a statistical technique based on the assumption that there is some commonality among the judgemental criteria such that they can be reduced to a small number of common dimensions or key components. A voluntary risk for example, usually correlates with 'controllable'. The two principal components to emerge from that analysis were labelled by the investigators as 'dread' and 'familiarity' (see Figure 8.1). Pesticides and food additives for example, are seen as unfamiliar but only moderately dreaded while alcohol is perceived as familiar and not dreaded.

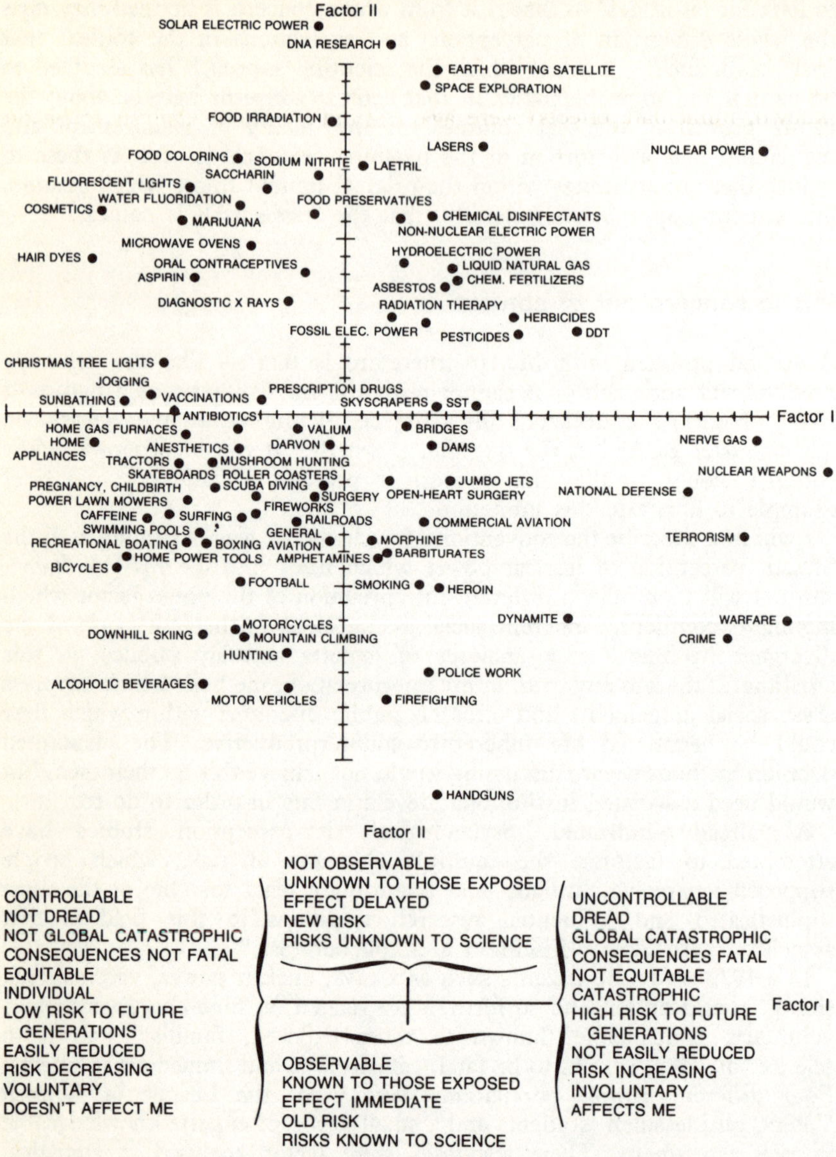

Figure 8.1 Hazard locations on Factors 1 and 2 of the three-dimensional structure from the interrelationships among 18 risk characteristics. Factor 3 (not shown) reflects the number of people exposed to the hazard and the degree of one's personal exposure. The diagram beneath the figure illustrates the characteristics that comprise the two factors.

Source: Slovic *et al* 1980.

Slovic and his collaborators repeated the study on a larger sample of college students with a list of some 80 hazards and 18 risk characteristics. The dread factor (made up of judgements such as catastrophic, uncontrollable, fatal consequences) and familiarity (for example, observable, known, immediate effects) were again found by factor analysis to be two independent orthogonal dimensions of risk perception to which all the other attributes could be reduced. This work has become probably the central paradigm in risk perception, the uncertainties in the whole process of factor analysis and psychometric observation notwithstanding.[20] It is therefore a good basis for exemplifying a sociological interpretation.

The orthogonal relationships of 'dread' and 'unfamiliarity' are shown in Figure 8.1, with the risk-activities located in this two-dimensional space. Fischhoff and colleagues called these respectively, the 'emotional' and 'cognitive' dimensions of risk-perception response.

The way in which this kind of work has been interpreted is that the key dimensions, 'dread' and 'unfamiliarity', are subjective psychological categories. They are thought to represent attributes associated in perception with objective physical properties of the technology concerned. Heavy negative weightings of these attributes then gives a highly negative 'utility' for a risk that might in terms of deaths and injury etc. be no greater or less than others. However, I suggest that they should be understood in sociological terms which leads to different practical interpretation of this work for policy.

The 'dread' dimension of nuclear power is widely associated with two particular symbols — the mushroom-cloud image of the atomic bomb and annihilation, and the unseen, all-penetrating image of radiation. Let us take the latter first.

The radiation dread-factor

The supposition that the invisibility of radiation carries an intrinsic sense of dread, is false and superficial. This is shown by examining Figure 8.2, taken from a newspaper supplement on nuclear energy in 1956. In those more trusting, paternalistic days, as the caption illustrates, the unseen nature of radiation was being socially constructed as a positive element of the image of nuclear power. It was part of the mystery which kept the controlling experts and elites suitably distanced, socially and intellectually, from a public which was evidently meant to simply look on in awe. It is striking in the 1950s the social distance and inaccessibility of nuclear power — its social invisibility — was part of the positive cultural legitimation of the technology. Thus social invisibility and radiation's physical invisibility were made to correspond as positive connotations. Now the reverse is true, for both the social and physical invisibility are no longer hidden. Thus this attribute is clearly only a code for a set of social relationships, and the association with one particular set (positive in the 1950s) or another (negative in the 1980s) is contextual, and socially achieved. Measuring its existence by psychological research is measuring that social achievement, not an intrinsic, individual mental state corresponding to a risk 'attribute'.

DOUNREAY —

to mankind will be the reward . . .

Millions of people, with only a glimmering of what the Atomic Age can mean to mankind, stand amazed at the fantastic prospect of heat, light and power issuing from a source that cannot even be seen. These people of the world, for whom the atom will be harnessed for service, gather their news of this great new power, day by day, from newspaper, radio and television.

These men and women can feel a sense of pride in the achievement of harnessing this power for peace.

Figure 8.2 Earlier perceptions of nuclear energy[21]

The extra dread 'caused' by the invisibility of radiation is not an inherent attribute either of radiation risks, or of public perceptions, therefore should not be incorporated in MAUA responses to public concern as if it were. It is a function of the social invisibility of the technology; ironically, more elaborate decision techniques may only make this invisibility worse.

The bomb dread-factor

Taking the 'atom bomb' part of the 'emotive' dread now, this can be seen as an oblique expression of an authentic empirical logic in the institutional structure of nuclear technology. There are two equally defensible, but contradictory rationalities in thinking about the possible connections between nuclear energy and nuclear weapons. One of these would say that the possibilities for horizontal proliferation of nuclear weapons (hence increased chances of atomic annihilation) by the spread of civil nuclear energy, are nullified by the Nuclear Non-Proliferation Treaty safeguards, managed by a competent International Atomic Energy Agency (IAEA). This is a legitimate (predictive and present) behavioural judgement. However an alternative judgement is that proliferation to unstable states is proceeding inexorably with the (not exclusive, but substantial) help of global civil nuclear energy diffusion. In this equally legitimate stance, safeguards are failing, and indeed it is unrealistic to expect that formal safeguards could ever control that proliferation. In this opposite view there is also a concrete empirical judgement about institutions, but this time one which creates an empirical association between nuclear energy and nuclear weapons. Thus the 'dread' association is not merely emotional, nor should it be judged as a (false) physical belief.

We therefore have two opposed beliefs about the 'dread' content of nuclear power, but they are not so much opposed physical claims as opposed social judgements. Which one of these conflicting views is 'correct' is beside the point in the present context. The point is that each one is equally empirically defensible and rational. To take the 'atom bomb' image of civil nuclear energy as a literalistic empirical assertion and to dismiss it as ignorant and irrational, is to miss the point entirely.

Thus the so-called irrational, emotive dread of the atomic bomb — the mushroom cloud as a risk perception multiplier for nuclear energy — can be taken as a symbolic language, containing a tacit behavioural judgement that the institutions supposedly keeping nuclear energy and nuclear weapons apart are ineffective. The objective scientific framework also makes an equally questionable social judgement, but recognises only direct physical factors (on which dimensions there is indeed no connection) and bemoans the emotiveness of those who claim the weapons' association. In effect however this is merely an authoritarian way of dismissing those underlying legitimate social judgements and concerns, and it only conceals the equally uncertain institutional judgement (that safeguards will work) underpinning its own 'objective' framework. Again, it is necessary to emphasise, this analysis does not take sides on the issues in dispute. It identifies a neglected dimension of the rationalities of all sides which needs to be properly addressed.

The psychological, 'scientific' approach to risk perception which analyses only in terms of individual mental factors such as dread, falsely obliterates these objective underlying institutional dimensions. It thus effectively consolidates the 'objective risk — subjective public perception' fallacy. In thus inadvertently dismissing alternative social judgements, the dominant framework tends to be provocative, and loses an opportunity to develop risk debates onto a more mature and constructive plane.

A parallel argument to that just given for the 'dread' dimension can be made for the 'unfamiliarity' dimension too. Differences between expert and lay perceptions of risks can frequently be resolved into differences of definitions of the scale and structures of the risk-generating system in question. The question of scale or structure may relate to judgements of social control. For example those who feel a technology is generally properly controlled will 'naturally' treat any proposal for a single plant in isolation from any associated developments. Conversely, those who feel generally excluded will be naturally concerned about what further expansions may follow the current proposal. Their risk problem is multiplied manifold. Notice also that their sense of 'dread' may thereby be amplified, suggesting non-independence of the orthogonal axes.

Apparently irrational differences of public perception of pesticide risks, between the large amounts used in domestic gardens, and the wish to ban the very same chemicals altogether in forestry or farming, can be explained in similar terms. One risk is seen to be under personal control, and reversible — the other is beyond control and may thus be assumed to go on for ever. Hence the definition of the risk-system is a function not only of the particular facility, or chemical, in question, but of that facility's or chemical's social structure of control. If that social structure is seen as alien, the risk-problem is open-ended, and logically so. The problem lies in the social relations as much as in 'public misperception' of the 'objective' risk-comparisons.

Conclusion

These sociological explanations are emphatically not intended to dismiss the realities of physical risks and damage (or their lack), nor to reduce physical beliefs to social relationships. Nor do they take sides in disputes between experts and laymen both need to develop and render more sensitive, their expressions of values and social commitments within their more concrete languages of risk.

I identify the interaction between public concern which focuses upon specific concrete risks and attributes, and that broader but no less real concern which underpins and shapes the more immediate and specific perceptions. The more diffuse concerns are sometimes seen by officialdom or experts as being unreal, and irrationally condensed onto symbolic victims such as the nuclear industry, or dioxin. Yet I suggest that they do have a real, objective and relevant grounding, especially when the full historical process of their germination and emergence is examined. Even if their basis is not recognised by their holders themselves, and even if they sometimes result in specific false estimates of risk magnitudes, they have to be addressed in these deeper terms by decision-making bodies. This does not mean they should be capitulated to, by asking people to put monetary value on their concern, or by asking others to do it for them. These deeper currents must be recognised by both 'sides', in enlarged institutional processes which can encourage their constructive expression, and mature negotiation. As I hope I have shown, enlarging the techniques does not in itself achieve this, and could even make it more difficult, unless the kind of institutional correlates of public risk perceptions I have identified analytically are properly recognised in practice.

What is meant by institutional innovations? It must be recognised straight away that no single formula is adequate to meet the variety of settings in which risk-management decisions are currently taken. Nonetheless some illustrations can be given. Many such decisions are determined in the United Kingdom via expert advisory committees. These usually provide no record of meetings, of how issues are defined, how diverse evidence is examined and judged and different disciplinary inputs reconciled. Their membership is often narrow in disciplinary terms, and they are not held to account for their judgements (except in very indirect ways via Parliament or the media if the issue is sufficiently controversial). All these could be changed. As another illustration, where risk-perception studies indicate that a common basis of concern is that a hazardous plant will be extended or multiplied in future, and that regulation is biased, lax or 'co-opted' by the industry, why not allow a community veto on any future development beyond the existing plan, and have locally accountable experts funded by the developer, with a remit to monitor, keep up with wider research on the relevant risks, and educate the local public? Other innovations could be appropriate for other settings.

In considering existing lack of the imagination in dealing with the institutional dimensions of risk perceptions, my own experience is relevant from working with a team of consultants during 1985 on a social impact assessment methodology for the Department of the Environment's BPEO study for radioactive wastes (see note 14). The question arose of where the issue of risk perceptions related to our terms of reference. I asked how in this context perceptions of different options for 'institutional packaging' of radioactive waste disposal might be dealt with, but it was not seen as at all relevant to the issue of public risk perceptions. Money compensation was the nearest that the discussion came to properly considering the interaction between risk perceptions and the institutional structures within which technologies and risks are delivered.

A basic message of risk perception research is that no decision, and no issue, is an island. One of the central methodological currents feeding decision analysis has been to ask hypothetical gambling questions to people. There are a variety of severe limitations about using this as a nominative policy framework for defining 'rational' responses to risks. One of the least attended to is that such artificial hypothetical decisions take no account whatever of the fundamental fact that people are involved in, developing relationships that have a history and, we hope, a future. Preserving trusted institutional frameworkers and social arenas is rationally a basic normative concern, and risks will be defined and managed with the health of such relationships centrally in mind. Decision analysis regards human interaction only as a means to ends, not as a value in its own right.

For all the laudable attempts to develop them, decision techniques have not as yet begun to address the underlying framework issues in public concern. Alongside practical attempts to extend them so as to incorporate public concerns we should also research different institutional options which may address and open those deeper framework questions. We should not delude ourselves that this will somehow achieve credibility and consensus where the conventional rational paradigm has failed. However the crucial point is that existing conflicts are essentially sterile and degenerative.[22] Their inherent intellectual and institutional structures discourage social and political learning; and all too often, short-term expediences in the search for legitimacy and credibility, give rise to longer-term mistrust and cynicism. We might at least aim for more authentic expressions of concerns and uncertainties. Granted that this will not end conflict, it could be more constructive and there may be positive trade-offs for the experts between less control (which is in any case increasingly illusory) and more credibility. Under present circumstances they risk losing both.

Notes

1. The most clearcut perspective here is the cultural anthropology of Douglas, Thompson and Rayner. See for example M. Douglas and A. Wildarsky, 1982, *Risk & Culture*, University of California Press, Berkeley; M. Thompson (this volume, Postscript); Rayner, S., 1986, 'Risk & Relativism in Science for Policy', in Johnson, B. and Covello, V. (eds), *The Social and Cultural Construction of Risk*, New York, Plenum, 5–23. See also Short, J., 1984, 'The Social Fabric at Risk: Towards the Social Transformation of Risk Analysis', *American Sociological Review*, **49**, and Wynne, B. 'Risk Perception, Decision Analysis and the Public Acceptance Problem', Chapter 11 in *Risk Management and Hazardous Wastes: Implementation and the Dialectics of Credibility*, Springer, Berlin, London, New York.
2. Starr, C., 1969, 'Social Benefit versus Technological Risk', *Science*, **165**, 1232–8, The critique by H. Otway has been neglected: Otway, H. and Cohen, J., 1975, 'Revealed Preferences: Comments on the Starr Benefit-Risk Relationship', IIASA Research Memorandum, RM–75–5, International Institute for Applied Systems analysis, Laxenburg, Austria.
3. See note 1.
4. Paul Valery, *History and Politics*, Harper and Row (English translation, 1962), p. 5.
5. Wynne, B., op. cit, note 1.
6. For different perspectives within the same general field, see: Raiffa, H., 1968, *Decision Analysis: Lectures on Choices under Uncertainty*, Addision-Wesley, Reading, Mass.; Keeney, R. and Raiffa, H., 1976, *Decision with Multiple Objectives: Preferences and Value Trade-offs*, Wiley, New York; Winterfeldt, D. von and Edwards, W., 1986, *Decision Analysis and Behavioural Research*, Cambridge University Press, New York; Merkhofer, M., 1987, *Decision Science and Social Risk Management*, Reidel, Boston; Sjoberg, L., Tyska, T. and Wise, J. (eds), *Human Decision Making*, Bodafors, Doxa; Fischhoff, B., Lichtenstein, S., Slovic, P., Derby, S. and Keeney, R., 1981, *Acceptable Risk*, Cambridge University Press, New York. For some case studies, see Keeney, R. and Nair, K., 1980, 'Selecting Nuclear Power Plant Sites in the Pacific North West Using Decision Analysis', in Bell, D., Keeney, R. and Raiffa, H. (eds), 1980, *Conflicting Objectives in Decision*, Wiley, New York; Lathrop, J.W. and Watson, S., 1982, 'Decision Analysis for the Evaluation of Risk in Nuclear

Waste Management', *Journal of the Operations Research Society*, **33**, 172–88. For an attempt to defend utility theory against criticisms, and to incorporate an element relating to (decision-maker) risk perceptions, see Raiffa, H., 1985, 'Back from Prospect Theory to Utility Theory', in Thompson, M., Wierzbycki, A. and Grauer, M. (eds), 1985, *Plural Rationalities and Interactive Decision Processes*, Springer, Berlin. For a brief overview, see Covello, V., 1987, 'Decision Analysis and Risk Management Decision Making: Issues and Methods', *Risk Analysis*, **7**, 131–9.

A fundamental problem not fully solved by multi-attribute methods is that the decision problem is still taken to be that of a 'decision-maker', reflecting the individualistic origins of this whole tradition. Thus its widespread application to public policy issues, where extra dimensions of conflict and interaction are involved, raises complex questions of how to define collective human processes. Many approaches and applications pretend to 'answer' these questions by appearing not even to recognise their existence.

7. Watson, S., Fischhoff, B. and Hope, C., 1984, 'Defining Risk', *Policy Sciences*, **17**, 193–210.
8. For examples, see Jones-Lee, M., Hammerton, M. and Philips, P., 1985, 'The Value of Safety: Results of a National Sample Survey', *Economic Journal*, **95**, 49–72.
9. Hetherington, J., 1975, 'Cost-benefit Analysis for Discharge Optimisation; Methodological Observations', Ministry of Agriculture, Fisheries and Food; Fisheries Radiobiological Laboratories, Lowestoft.
10. These were analysed at a meeting between representatives of the Health and Safety Executive, National Radiological Protection Board, and Risk Working Group of the Economic and Social Research Council, in London, October, 1985. 'Willingness-to-pay' or 'expressed preference' surveys were criticised by Risk Working group members as too abstract and unreal, leaving too many uncontrolled variables to allow valid practical interpretation into 'acceptable risk' decisions. The HSE and NRPB went ahead with a large-scale survey nevertheless; its results were not very clearcut.
11. Webb, Geoff, 1985, National Radiological Protection Board, Harwell, UK, paper to the HSE, NRPB, ESRC meeting, London, October.
12. Ibid, and follow-up letter to Jennifer Brown of ESRC Risk Group, December 1985.
13. On motorways, the Leitch Report was critical of BCA methods. *Report of the Advisory Committee on Trunk Road Assessment* (Chairman G. Leitch) London, HMSO, 1978; on the Third London Airport, Roskill Inquiry's use of BCA, see Adams, J.G., 1971, 'From TLA to Airstrip One', *Geographic Journal*, **137**, 468–504.
14. Assessment of the Best Practicable Environmental Options (BPEOS) for the Management of Low- and Intermediate-Level Solid Radioactive Wastes', Department of the Environment, Radioactive Waste Division, HMSO, London.
15. Watson *et al*, op. cit, note 7.
16. In the psychometric paradigm of Slovic, Fischoff and co-workers. See for example Slovic, P., 1987, 'Perception of Risk', *Science*, **236**, 280–5; Slovic, P., Fischhoff, B., and Lichtenstein, S., 1985, 'Characterising Perceived Risk', in Kates, R.W. *et al* (eds), *Perilous Progress: Technology as Hazard*, Westview Press, Boulder, 91–123, 1984, 'Modelling the Societal Impact of Fatal Accidents', *Management Science*, **30**, 1984, 464–74; 1980, 'Rating the Risks', *Environment*, **21**, 14–20, 36–39. 'Perceived Risk', in Schwing, R.C. and Albers,

W.A. (eds), 1980, *Societal Risk Assessment: How Safe is Safe Enough?* Plenum, New York. 1987, 'Ripples in a Pond: Forecasting Industrial Crises', *Industrial Crisis Quarterly*, **1**, Winter, 34–43.

17. See Wynne, B., op. cit, note 1 for a fuller explanation of this argument.
18. See Wynne, B., 1988, 'Establishing the Rules of Laws: Constructing Expert Authority', in Smith, R. and Wynne, B. (eds), *Expert Evidence: Interpreting Science in the Law*, Routledge, London and New York.
19. See note 16.
20. For some cogent questions about the psychometric paradigm, see Otway, H. and Thomas, K., 1982, 'Reflections on Risk Analysis and Policy', *Risk analysis*, **2**, 69–82. As a non-psychologist, I do not wish to engage in methodological criticisms of this approach. My reinterpretation is to suggest not that the work is wrong, but that it is only part of the full story — indeed many of Slovic *et alia's* attributes, as they recognise, involve risk management rather than physical risk *per se*.
21. 'Atomic Power', supplement to *Financial Times*, London, 15 October 1956, on occasion of the opening of the United Kingdom's first electricity generating nuclear power station at Calder Hall, October 1956. 'Dounreay' refers to the prototype fast reactor facility which was built in 1951 in Caithness, Scotland.
22. For some constructive proposals for reorientation, see Rayner, S. and Canter, R., 1987, 'How Fair is Safe Enough? The Cultural Approach to Societal Technology Choice', *Risk Analysis*, **7**, 3–9. Notice we are now talking about the politics of technology and not 'the' rationality of risk perception. That seems an appropriate and necessary shift of focus.

Chapter 9
Conclusion and Policy Perspectives

Jerry Ravetz, Silvio Funtowicz and Alan Irwin

The essays collected in this book are the outcome of particular research projects conducted by the authors, guided by their own ongoing programmes and for which the resources were secured by their own efforts. They show a unity of theme and approach, partly because of the discussions of the Risk Working Group which preceded the presentation of some of them at the joint ESRC and DOE seminar, but more because of the sense of common purpose shared by all those who worked on the risks group. This could best be expressed as an awareness of the need for theoretical tools of the sort that can be supplied only by one or another field of social study, but also a commitment to use those tools in the solution of practical problems rather than as objects of purely intellectual concern.

In such a collection of reports on special studies, there is a tendency for the unproblematic background that is known by all the authors to be treated as known. This enables their own arguments to flow smoothly, but it may leave the reader wondering about the source of their ideas and the context in related sorts of studies. To avoid this, Jennifer Brown prepared an introductory overview which provides a setting for the present essays, in the scholarly work on risks from a variety of approaches. Her survey includes the background in the growing salience of environmental risks, the variety of approaches within the behavioural sciences and the methods by which risks (and the perceptions of them) are assessed. The topic of 'technical risk assessment' receives its due attention, but anyone who has considered that that is central to the task and all the other aspects peripheral will be impressed by the sheer bulk and detail of the other, complementary studies.

A similar point is established by Brian Wynne in the first of the five essays on 'approaches'. He reviews a set of case studies covering a wide variety of situations and events. He shows by example that 'expert' and public perceptions of risks share the same property, of being a partial framing of a risk assessment problem, depending on particular roles and experiences. In the case of 2,4,5-T, farmworkers' representatives introduced anecdotal evidence from field experience to cast doubt on the reliability of safety procedures, while the PAC scientists based their judgements on laboratory tests which proved that (under proper use) the substance must be harmless. Similar considerations explain divergences of policy prescriptions in the cases of animal hormones and toxic waste landfills. A rather special sort of perception of a risk assessment problem explains a report by the Health and Safety Executive (HSE) on the Abbeystead explosion; they exculpated all concerned, while evidence of negligence was accumulating on all sides. It seems likely that their study was designed for advice on bringing a criminal prosecution, where criteria of proof are very strict; and being released without explanation, it gave a misleading impression of the HSE's general approach to the control of risks. Can all parties to risk debates, including experts, recognise the limitations of their own frameworks? Social scientists can help in this new task.

The case of 2,4,5-T is reviewed more extensively by Alan Irwin; he analyses the different sorts of 'proof' that were involved, and also the different sorts of evidence that would be appropriate for each case. The issue can be slightly simplified as one of 'burden of proof' — is the situation deemed safe, or dangerous, until proved otherwise? A somewhat paradoxical issue is the topic of his other case study: the debate on improvements in road safety consequent on the adoption of compulsory seat-belts. No one denies that, given a collision, a person in a car is safer with a belt. But John Adams argued that there is no statistical evidence of an improvement in road safety after compulsory seat-belts, and made the hypothesis of a 'risk compensation' by which drivers tend to increase their risks in other respects. There was an angry reaction from the experts, and a breakdown of communication; this may have been a result of the commitment of all sides to saving lives, and also the immediate policy implications of the issue.

Sometimes the 'experts' disagree among themselves, or even take time to enquire about a potential risk while the affected public awaits their conclusions. This is what happened at Seascale, near Sellafield, when a Yorkshire TV programme raised the alarm about an excess of cases of leukaemia in the area. The only known cause of leukaemia at that time was radioactivity; and the BNFL reprocessing plant had already had its share of malfunctions and suspicions of excess emissions. Here the reaction of the public could not be NIMBY (Not In My Back Yard), since Sellafield depends on BNFL for its very existence. The study of 'risk perception' in such a conflicted case requires a flexible and sensitive approach, so that the contradictions between groups and even within individuals can be elicited and tactfully portrayed.

Usually environmental risks are perceived only after they have become a serious threat to some interest; and the creation of a regulatory system is then fraught with difficulties. Sometimes, however, it is possible for society to see a risk being born; such is the case with genetic engineering, either in large-scale industrial operations or (more crucially) with planned environmental release. Jennifer Brown shows that in the United Kingdom public opinion is not aroused on this issue, but there is already a sense of concern, which could easily become hostility if the industry and its regulators do not maintain public confidence. Jerry Ravetz analyses the technical problems of risk assessment in this field, showing the extreme difficulty of modelling the behaviour of information carried on biological systems into the natural environment. He mentions the 'fine-tuning paradox': engineered species released into the environment must be sufficiently hardy to survive long enough to do their job, but not so vigorous as to upset the local eco-system; and they must be known in advance, with good certainty, to have this finely balanced property. One solution to the assessment problem is to cultivate 'ecological wisdom', making a new sort of 'natural history' popular as a necessary complement to the laboratory science now dominant in biology.

It is clear from all these studies that 'the public' that reacts to risks is not uniform in its composition or in its attitudes. Indeed, there are so many risks around us all the time, that we must, individually and collectively, select those which are most salient for us at any time. An understanding of the criteria for selection would be of great use in the management of risks about which there is public concern. Peter James and Michael Thompson describe the 'plural rationality approach', which applies the 'grid/group' methodology of the anthropologist Mary Douglas. This provides a four-fold classification of social influences on individual behaviour, by the degree of constraint and of belonging, respectively. The idealised types are those of hierarchical societies, individualistic networks, egalitarian groups and atomised individuals; their rationalities are (respectively) bureaucratic, substantive, critical and fatalistic. In these terms we can explain reactions to nuclear power, and even to cancer.

Passing now to 'tools' we have two studies of attempts to provide a formal or quantitative analysis for the management of risks. Brian Wynne reminds us of the inevitably value-laden character of any quantification, or even classification, of risks. The simplest approach, Benefit-Cost Analysis, necessarily reduces all values to a single quantitative measure, while Multi-Attribute Utility Analysis at least enables the explicit display of different dimensions of a perceived risk problem. Even this presupposes a single, context-free decision problem, which improves an inflexible value-framework around the choices which are allowed. He reminds us of the importance of attributes which are usually seen as psychological, but are actually sociological: 'unfamiliarity' and 'dread'. The combination of these two has made nuclear power uniquely vulnerable to public distrust, relating more to the history and institutional style of the industry than to probabilistic risk analysis.

Concentrating on the uncertainty that is endemic to risk management, Silvio Funtowicz and Jerry Ravetz have devised methods for its management. They call their system NUSAP, an acronym for the five places in the notational system, Numeral, Unit, Spread, Assessment and Pedigree. The last three of these describe inexactness, unreliability and boundary with ignorance, respectively; these are three major sources of uncertainty on the cognitive side. While Spread and Assessment have their analogues in statistical practice, Pedigree is novel: it is an evaluative account of the information, related to the conditions of its production or perhaps of use.

As a matrix whose columns are roughy normative entries, it provides a portrayal of the 'state of the art' of the field in which the information was produced. By means of NUSAP, uncertainties in quantitative information are clearly displayed, and the task of communication and quality-control are facilitated.

What policy perspectives can be drawn from these various approaches and tools? The first can be expressed negatively: that the management of technological and environmental risks is not, and cannot be, the simple application of a mathematical science. There is a corollary to this conclusion: any institutions responsible for managing risks which maintain that their science is definitive, and that all who doubt it are afflicted by ignorance or emotion, will themselves be at risk of a loss of credibility. Their policies will be judged even less on their scientific grounds, and more on a negative perception of their institutional behaviour.

The lessons from our studies are not, however, all negative. We have seen that there is a complementarity between the risks as existing out there, possibilities of harm, and the risks as *both* created and perceived in particular social contexts. The call for an involvement of 'the public' in the management of risks is not an expression of democratic or populist sentiments. For each risk has a set of publics, which are affected by the risk as it is created, and which in their turn react back on the institutions responsible for it. The interaction is there as a matter of fact, technical and political. The question is whether it will be well managed, or otherwise.

What emerges from our studies is the opening of a new dimension to the activity of citizens in their communities. Each risk problem is complex, affecting different people in different ways and involving several sorts of expertise, perhaps conflicting among themselves. Ideally there should not be a simple polarisation between remote institutions who impose a risk and embattled communities who oppose it. If a constructive dialogue is to be fostered, so that risks can be properly managed, then some of the lessons drawn from our studies may be useful. These are in the nature of recommendations which of course will require serious research and careful experimentation before they can be implemented with confidence.

The first point is that new forms of citizen involvement require new structures for their accomplishment. There are already a variety of tribunals in existence, the various sorts of enquiries and also (in some extreme cases) the civil or criminal courts. In general these were designed around particular aspects of risks problems, or perhaps (as in the case of planning enquiries) quite a different problem altogether. None of them have citizen participation built into their structure. Even when this occurs, there is no constitutional provision for enabling that side to participate in a competent and hence responsible fashion. The provision of resources, in the way of exertise and expenses, has been considered on some occasions, but never adopted as a principle. In the absence of such support, local and community interests will always feel themselves excluded in practice if not in formality from the real decisions; and they cannot then be blamed if they assess the risk in terms of an array of institutions perceived as uncaring and arrogant.

Part of the process of sharing in the management of risks will have to involve a changing awareness of what that process is all about. If there is not a once-for-all conclusion about 'acceptability' based on scientific facts, then the risk will be seen as something that evolves along with the installation that creates it (from plan, through design, construction, start-up, operation, to ageing and disposal). Each phase presents different problems and possibilities to the various interests involved with the risk. A continuous, ongoing process is appropriate here, where all the different parties learn about the risk, each other and themselves, through the cycle of development of their common problem.

Out of such a continuous process of dialogue there could also emerge a better mutual appreciation of perspectives, between those with a general scientific expertise and those with a local, personal experience. This would be enhanced by a programme of education on risks, now established in various courses in secondary education but not yet systematically in the more traditional subjects. Following this we could imagine centres where science as applied to risks and to other problems of everyday life could be made available through advice to enquirers.

All this brings us a long way from the idea of a mathematical science of risks, the possession of qualified experts, whose conclusions about acceptability of risks should be believed by any unbiased member of the public. The world is less tidy than that, but also more interesting. In his Postscript, Michael Thompson draws out these conclusions in a lesson for experts, publics and scholars alike.

Postscript

Engineering and Anthropology: Is There a Difference?

Michael Thompson

Risk and uncertainty: a scientific approach

Risk assessment — 'The new science of risk assessment' (Silcock 1979), as it has been hailed — sets out to do three things: to *quantify* the dangers of modern technology, to *compare* them, and to *devise rational policies* for dealing with them. Though the techniques of risk assessment are mathematical and at times complex, its conclusions are simple enough and are often displayed in the form of *risk tables* for a variety of activities: coal-mining, rock-climbing, hang-gliding, and so on, each with its appropriate odds, for example: 'The chances of being killed rock-climbing are 1 in 13,000 in any one year'. Once the risks have been quantified and compared in this way, it is but a short step to marry risk assessment to cost-benefit analysis (and a particular sort of engineering to a particular sort of economics) to produce *risk benefit analysis*, the technique that, among other things, places an exact money value on a human life: US $287,175 (Hapgood 1979). Two stages are involved: first, determining the risks of various accidents within each technology or activity and, second, making the comparisons and drawing the policy conclusions from them.

The basic aim of risk assessment, from the beginning, has been to find out what the risks really are. After more than a decade of trying, the main conclusion of all the research activity that this 'new science' has stimulated is that this aim is unattainable: 'A distinctive characteristic of our natural world is that it typically is not and cannot be known to the desired degree of precision' (Slovic, Fischhoff and Lichtenstein 1979, 370). Risk assessment, in other words, has set out to achieve the unachievable. Risk assessment is an heroic, but hopeless, enterprise: an intellectual charge of the light brigade. *C'est magnifique mais ce n'est pas la science.*

The question now arises: what would a scientific approach to risk look like?

Ten searching questions

The argument in favour of risk assessment runs something like this:

Method: 'The simplest way to put the risks of big technology into perspective is to compare them with the risks we run in everyday life' (Silcock 1979).

Application: 'In any gamble it is foolish not to find the odds since these are the only rational means we have of taking vital decisions, not only about personal safety but also about national policy' (Silcock 1979).

Before accepting these seemingly reasonable and innocuous claims we should ask the following questions:

(1) Can risk assessment tell us what the risks in a big technology, such as nuclear power, really are?
(2) Is the comparison with other risks valid and meaningful?
(3) If we cannot find out what the odds are does it follow that we cannot act rationally?
(4) If the answer to question 3 is 'no', what are these other rational modes of decision-taking?
(5) If the real odds (assuming they can be determined) are not matched by individuals' perceptions of the odds, should they override individual values in arriving at the best social choice?
(6) Since risk assessment has only existed for ten years or so, we know that social choices can be arrived at without this ruthless, direct comparison of risks, but we do not know whether the reverse is true. That is, we do not know whether social choice is possible with total comparability. To put the question at its starkest: are risk assessment and democracy inimical?

Then we should query some of the basic, but implicit, assumptions in risk assessment:

(7) That all risk is nasty.
(8) That it is always more desirable to know the odds than not to know the odds.
(9) That there is a difference in kind (i.e. a non-comparability) between just two categories of risk: voluntary and involuntary.

Finally we should ask this question:

(10) If risk assessment is only one out of several rational modes of decision-taking, why should it be that some individuals and institutions should be so attracted to it as to believe not only that it is the best, but that it is the only mode?

Of course, as anyone familiar with the risk debate will point out, each of these ten questions has been asked, and answered, by someone, somewhere. Taken together they provide, not a new point of departure, but a convenient means of summarising all the serious criticisms that, over the years, have been levelled at risk assessment. But they do much more than simply list the shortcomings of risk assessment; they turn the tables on it. Though these criticisms have been seen by the self-professed hard scientists as the negative and unhelpful carpings of radically motivated softies, the fact is that these carpings, put together, *are* the scientific approach to risk.

So these 10 questions are a research programme: a strategy for establishing a coherent, usable and scientific framework for guiding decision-making on technology. Since each of these questions has already been asked and answered, the task we face is not so much one of creation as of assembly. Even so, it is a considerable undertaking and not one that can be achieved with any measure of conviction and comprehension in a short chapter such as this. I will restrict myself here to conveying the feel — the essential character — of the scientific approach to risk by exploring the answers to just two of these questions: numbers 1 and 3.

Question 1: Can risk assessment tell us what the risks in a big technology really are?

Leaving aside the philosophical objections that can be raised over the odds that a penny will come down 'heads' when you have called 'tails' (Keynes 1973), we should look carefully at what is involved in the extension of probabilistic reasoning from coin-tossing to high technology. As you go from the flip of a penny to the melt-down of a reactor core so you go from a closed system to an open system: from science to engineering. The engineer, of course, does use a lot of science but, unlike the experimental scientist, he does not work on the laboratory bench; he works in the world. His skill, his judgement and his art lie not so much in his mastery of the science he is using as in his handling of the inescapable tension between the closed-system assumptions of that science (the ideal gas, the black body, the coin that cannot land on its edge) and the open-system reality (the possibility of unconsidered and unconsiderable outcomes) of the world in which he is working. That is what engineering is. That, as engineers well know, is the source of their dignity: the unique foundation on which their entire profession is built (Chilver 1975; Muir Wood 1980).

This is an absurdly simple and elementary distinction — first year (first day, even) engineering course stuff. Nor is it a distinction drawn from the outside, by social scientists or philosophers. It is an engineering distinction; it is the engineer's own understanding of what he is. All I am going to do from the outside, as it were, is take this distinction and see how risk assessment measures up to it. Even then, I will rely heavily on those on the inside — engineers — to guide me.

Octavius Critchley (an engineer and a member of the Nuclear Installations Inspectorate of the British Health and Safety Executive) has developed this closed system/open system argument clearly and forcibly and has asked why the inappropriate quantitative methodology of the bench scientist should increasingly be applied to nuclear installations, especially when the appropriate qualitative methodology of the engineer that was employed until recently has been so successful that 'no one has been killed nor has anyone been injured by radiation on any nuclear site'. He goes on to provide the answer:

> The Rasmussen (1975) Safety Study (which was based on *quantitative* safety analysis) assesses the accident risks for a large atomic power programme at orders less than that to which people in the USA are exposed in normal living. Though it specifically avoids making any recommendations, a decision maker would conclude on scientific grounds that nuclear power is an acceptable risk . . . interest in this kind of prophecy has increased greatly and diverse methods of prediction and especially systems analysis have flourished under the stimulus and encouragement of politicians and administrators. The apparent reduction of dilemmas to a choice between numerical values is attractive because it enables proximate decision makers to compensate for lack of knowledge of the disciplines involved and this is especially so in technology. [Critchley 1978]

If Critchley is right, then we have here an example of the replacement of an appropriate methodology by an inappropriate one as a result of political and organisational pressures. Could it be that we are getting, not the best available management of technological risk, but the most administratively convenient one? To ask this question is to open up an altogether different approach to risk — an approach that focuses not so much on the risks themselves as on what they are being used for.

What kind of social being, we should now ask ourselves, sets a high value on administrative convenience? And what kind of a world would furnish him with a repertoire of risks conveniently arranged for him to administer? The answer to the first question is: someone in the upper reaches of a large hierarchical organisation. And the answer to the second question is: a world that is both finite and fixable. We can then go on and say that such a social being is behaving rationally if he insists (often largely at the behest of his unquestioned assumptions — his *a priori*) that that is how the world is.

Now, by introducing the notion of world-making, of socially malleable *ideas of nature* (M. Thompson 1988), we have uncovered a plausible rationale for the risk assessor's seemingly irrational urge to attempt the impossible. We also have a plausible explanation both for his resistance to the 'uncertainty principle' inherent in the open-system view of risk and for the technocratic tenacity with which he clings to his closed-system/bench science account of it all. Completeness, closure, clarity, compartmentalisation and control: these are the great claims that stabilise and justify hierarchical organisation. If a hierarchical organisation aspires to handle technological risk then it will be predisposed (for the soundest of organisational reasons) towards accepting and insisting on three things: first, a closed-system view of the world within which all risks are discoverable and measurable; second, a fixable nature within which those with the requisite skill and expertise can exercise control and; third, a definition of scientific rationality that emphasises specialisation, predictability, quantification and certainty-creation.

This, of course, is only one socially induced rationality and we should now press on to enquire what others can, and do, exist and what kinds of social organisation they variously help to stabilise. The idea of this approach is to tease out the various *plural rationalities* (Grauer, Thompson and Wierzbicki 1975) not with a view to determining which one is right (that is a meaningless question) but, rather, with the view that each will be appropriate in certain, specifiable circumstances and inappropriate in others. As we move from rightness to appropriateness so the variety and contradiction of management modes cease to be an obstacle to the implementation of rational policies and become a valuable resource: perhaps the ultimate resource. To harness this resource we have to overcome the narrow tyranny of the particular rationality that, as Critchley rightly fears, is currently driving out other, often more appropriate rationalities from the whole field of technology assessment and risk management. To do this we must query that orientation's central assumption: that quantification is a pre-condition for scientific rationality.

Question 3. If we cannot find out what the odds are does it follow that we cannot act rationally?

Risks can entail four sorts of uncertainty. First, when we toss a coin we are uncertain whether it will come down 'heads' but (certain philosophical provisos apart) at least we know what the odds are that it will do so. Second, we know that a particular accident can occur (in a nuclear power station, say) but we are uncertain as to its likelihood. Third, we may be uncertain as to whether something that we know can happen is dangerous or not. Fourth, we may not even be aware of the existence of that something that may or may not be dangerous.

Within economics the first two correspond to the celebrated distinction between 'risk' and 'uncertainty' (Knight 1921), while the last two are either ignored or uncritically subsumed under 'uncertainty'. Engineering is rather more discerning. There, the first kind of uncertainty is handled by experiment. In an experiment there is a complete theory predicting the behaviour of the system and the system itself can be isolated (put on the experimenter's bench, as it were), subjected to certain carefully controlled changes, and its behaviour observed to see whether it is consistent with the predictions of the theory. The second kind of uncertainty is handled by test. In this case there is no complete theory but the system itself can be isolated and put through a realistic series of changes to see what happens to it. The third kind of uncertainty (where even testing is not feasible) is handled, first, by conceding that it exists and, second, by drawing upon a wealth of shared experience in the handling of that sort of thing. The notions of the maximum credible accident (MCA) and the maximum hypothetical accident (MHA) and the way they are then related to the design criteria within the site and to the selection criteria for the site itself, respectively, provide an example of how this sort of uncertainty has been handled. The last kind of uncertainty calls for the full exercise of the engineer's open-systems art. 'Cockcroft's Folly' — the vents on the top of the Windscale piles that averted a major accident of a type that, at the time of construction, was not conceivable — is an example of this sort of 'sound engineering judgement'.

Bench science, by contrast, restricts itself to the experiment, and this can lead to some interesting confusions. The Rasmussen Reactor Safety Study, for instance, adopts this bench science approach for the construction of its vast fault tree within which each box (each possible fault) is given a numerical probability. Once you have such a tree you can derive, by impeccable logic, the probabilities of particular accidents that can result from particular concatenations of faults. All you have to do is multiply the probabilities in the boxes along the particular sequences of faults that can lead to those particular accidents. But the engineer is reluctant to do this, preferring instead the woolliness of his 'sound engineering judgement' to the exercise of a clear-cut and indisputable logic. Now the incomprehension, and exasperation, start.

'Do you disagree with the probabilities?' the risk assessor asks. 'No' replies the engineer. 'Do you disagree with the fault tree — with the chains of events that can lead to certain accidents?' the risk assessor asks. 'No', replies the engineer. 'Then how can you possibly object to the multiplying of the probabilities?' demands the risk assessor. 'Oh, I don't object to all such multiplications,' says the engineer, 'just this one here, and this one . . . and that one there', pointing to particular branches within the fault tree. This is too much for the risk assessor. To him, a tree is a tree is a tree. 'How', he cries, 'can you possibly justify accepting multiplication along some branches and not along others?'

'Ah, its perfectly simple', says the engineer. 'See that box there, labelled "can-splitting" — its probability was arrived at by experiment. That box there, labelled "can-popping" — its probability was arrived at by a series of tests. And that box there, something to do with a possible pipe system failure — its probability was arrived at by getting a number of leading experts together, giving them a good lunch, and then asking them for their best estimates and taking the average. I would accept the multiplication of probabilities arrived at by experiment. I would accept the individual probabilities arrived at by test, but not their multiplication, and I would neither multiply nor accept probabilities of the third kind, preferring instead to use these numbers as just a rough (but still valuable) qualitative guide. And then, of course, I would note that the fault tree does not allow for Irish labourers urinating into the housings for the post-stressed reinforcing cables of the containment vessel whilst the plant was under construction, and I would like to be able to take that sort of thing into account as well.'

Most of the labourers on these construction sites are Irish. At tea and meal breaks they consume considerable quantities of tea (and sometimes beer) and are reluctant, when nature has taken its course, to make the laborious and time-wasting descent from their lofty scaffold. Since their Hibernian modesty precludes the alternative adopted by some of their younger (and more exhibitionist) English colleagues, they make discreet use of the concrete openings so conveniently positioned along their work station. Such is the structure of authority and control on the site that the information concerning the ultimate purpose of these openings — information that, if they possessed it, might close this option to them — cannot reach them.

Where does all this leave the risk assessor (not to mention his inexpert client, the decision-maker?) His prior commitment to a closed-system/bench-science approach leaves only one strategic avenue open to him. His sole aim has to be to convert all this uncertainty to the coin-tossing kind — to determine what the risks for each box really are — and his justification for this inescapable strategic commitment is the assertion that if we do not know what the odds are we do not have any 'rational means of taking vital decisions, not only about personal safety, but also about national policy. Such certainty-creation would be fine as a strategy if everything could be made certain. Alas, if there is one thing of which we can be certain it is that we cannot be certain of everything.

What are we to do about all that uncertainty that remains, even after the risk assessor has calculated every risk he can calculate? Well, for a start, we can ask whether it is indeed true that people cannot act rationally in the face of incalculable risks. The answer (hardly surprisingly, since probability was only invented a few centuries ago, while people have been taking risky decisions for millennia) is no. People can, and do, act rationally when confronted by such risks. What is more (and this is where the political culture of risk comes into the picture) their rational actions sometimes deliberately impose incalculability upon risks that are perfectly calculable. That is, they deliberately and rationally follow a course of action that runs exactly counter to the one which the risk-assessor has to insist is the only rational one!

Myopic opportunism and its far-sighted rival; two risk handling rationalities and their interaction

'I say, I say, I say', cries the omniscent and unstoppable music hall comedian. 'I don't wish to know that', protests his ineffetual stoodge. As with art so with life. To many happy-go-lucky individuals there is something inherently distasteful about the risk assessor's relentless pursuit of the risks of everything and, sensing this, they begin setting up the barricades, laying the false trails and putting down the smokescreens:

> People who claim to know jackrabbits will tell you that they are primarily motivated by fear, stupidity and craziness. But I have spent enough time in jackrabbit country to know that most of them lead pretty dull lives; they are bored with their daily routines: eat, fuck, sleep, hop around a bush now & [sic] then . . . no wonder some of them drift over the line into cheap thrills once in a while; there has to be a powerful adrenelin rush in crouching by the side of the road, waiting for the next set of headlights to come along, then streaking out of the bushes with split-second timing and making it across to the other side just inches in front of the speeding wheels. [H. Thompson 1980, 206]

Hunter Thompson is sending up the straights by outrageously acting the part of the anthropologist who, having done his fieldwork among the jackrabbits, actually understands the rationale behind their seemingly irrational behaviour.

And he is right. His little bit of fun-poking reveals two impressively contradictory rationalities — that of the jackrabbit (as interpreted by the anthropologist who has lived among them) and that of the people who 'claim to know jackrabbits'. But there is an interesting asymmetry in their descriptions. The anthropologist is concerned simply to establish the legitimacy of the jackrabbits' rationality, while those who claim to know jackrabbits are concerned to impose their own rationality on others, and this they do not just by pointing out the legitimacy of what they themselves do but by going one step further and denying the legitimacy of what others (jackrabbits, among them) do. The first is prepared to countenance a plurality of rationalities: the second is not. The first is tolerant of contradictions; the second is not, and it sustains itself (and advances its imperialist aims) by making accusations of irrationality against the others.

The study of witchcraft got nowhere until anthropologists stopped looking at witchcraft itself and looked instead at witchcraft accusations. Much the same, I suspect, holds for irrationality. While psychologists and philosophers argue endlessly over whether or not human irrationality can ever be demonstrated, the irrationality accusations and their revealing patterns (Lord Rothschild's (1978) intemperate remarks about 'eco-freaks and eco-nuts', for instance, and the no-less intemperate responses of the greens) are being laid out for any who have eyes to see them. Just map the small number of directions in which the various accusing fingers are pointed and the offending rationalities (and their institutional origins) are soon revealed in all their antagonistic splendour.

Himalayan mountaineers, for instance — both Europeans and Sherpas — follow the jackrabbit strategy of rejecting readily available information about the long term so as to be able to concentrate opportunistically on just those few (but large) risks that swim into their short-term ken (M. Thompson 1980). When this strategy is related to the sorts of individualised contexts in which they operate, the behaviour of these mountaineers makes a good deal of sense. We may not approve of what they are doing, of course, but there *is* a rationale behind it — a rationale that is based on the shrewd and daring assessment of the risks they face in the here-and-now, not in terms of the readily available historical data for the activity they are engaged in, but in terms of their own skill, experience and judgement. In other words, they go to considerable institutionalised lengths (these strategies are rooted in enduring patterns of social relations, rather than in the individuals' personalities) not to know the odds they face.

Set against this, the risk assessor's insistence that rational decisions are possible only when you know the odds begins to look more like an irrationality accusation than a statement of fact. One of the best ways of stopping people from doing something you do not want them to do is, first, to state authoritatively that it is impossible to do it and, second, to label as deviant (irrational) those who nevertheless attempt to do it. Is this what the risk assessors and the hazard managers, unwittingly and from the best of intentions, are doing to the rest of us? (Clark 1980).

At this point in the argument the exasperated risk assessor will usually enquire whether I seriously expect anyone to believe that the best way to manage the risks in high technology is to close our eyes to them. Of course, I do not. All I am saying is that his strategy of certainty-creation is only one of the socially available ways of handling risk.

My argument is that we should transcend the narrow provincialism that closes in the moment we demand to know which is the right way, and that we should begin, coolly and dispassionately, by discovering what these different risk handling styles are. Since each has been around for generations, and since each contradicts all the others, there must be something in it; it must be effective in certain circumstances; if it wasn't it wouldn't still be around. Each, therefore, has something to contribute. What we need, therefore, are two things: a lexicon of risk handling styles and a set of selection criteria that will help us to understand (often against our own provincial convictions) just when and where each is most appropriate, and just when and where each is least appropriate. That I submit, is the proper (and achievable) aim of a scientific approach to risk.

Engineering as tacit anthropology

So I am arguing for an anthropology of risk — hardly surprising, since I am an anthropologist. What may be surprising, though, is my idea about where we should start looking for such an anthropology. We could, of course, look at the burgeoning literatures on the cultural theory of risk, on the sociological science, on the social construction of reality and so on, but I prefer not to. I have no objection to people looking in these places (I have done a lot of it myself), it is just that I do not think this is the best way for social scientists to communicate with engineers, civil servants, policy makers, industrialists, environmentalists and the like; with doers rather than thinkers.

Nor is it just a matter of public relations. In many ways the practitioners (engineers and lawyers, for instance) are way ahead of the social scientists. The fact that all too easily gets overlooked is that, over the centuries, their theories and practices (unlike those of the social scientists) have been hammered out on the anvil of implementation: the criterion of acceptance is not simply that what the professional proposes is consistent and defensible but that it works as well. Or, to be more precise (and this is where the anthropology comes in), the criterion is that what they do is perceived as working well by those who, if they perceived it to be working badly, might react in such a way as to undermine the credibility — the public standing and esteem — of the professional. Professions, to be successful, have to be socially viable; they have to be mediating institutions; they have to contain within themselves a variety of approach and definition sufficient to match that which they will encounter among their diverse clients — us. The result is that, though such professions may not be able to provide explicit descriptions of all the various rationalities and risk-handling styles that they continually bump into, they none the less do take them into account. They have, as we say in the social sciences, 'internalised' them.

The role I am advocating for the anthropologist is a modest and, I hope, temporary one. His present task is to get under the skin of the engineer and draw out (we would say 'de-construct') this internalised understanding.

And the reason why he needs to do this now is that the environment of the engineer is changing so fast that internalisation is simply not enough. The engineer, as he passes through the turbulent transition from industrial to post-industrial society, needs a much more explicit understanding of all the artful things he does; he needs a greatly enhanced capacity for self-reflection. The anthropologist's temporary role, therefore, is to provide and polish the looking glass. Or, to put it another way, to give a more explicit and precise account of the nature of engineering judgement so that engineers can more readily resist those forces that are at present acting in such a way as to destroy that judgement.

The turbulent transition

Civil engineering is defined as the harnessing of the great forces in nature for the benefit of mankind. Though electricial and mechanical engineers sometimes chide their civil siblings for not having altered their ways since they put up the pyramids, there is still plenty of mileage left in this ancient definition; all it needs, if it is to see them through the next few millennia, is a slight shift in emphasis. Historically the benefits have usually been so obvious, and the technical obstacles so challenging, that the engineer's attention has (quite rightly) been focused on just two of the elements that make up his definition — the science that will enable him to understand the great forces and the engineering art that will allow him to harness them safely. All that has now changed. The really knotty problem now is to know what constitutes a benefit. The engineer used not to have to bother himself too much about that question; now he has to. That, in short, is his transitional predicament.

Buckminster Fuller was fond of pointing out that Edison did not need a licence to light up the world. However, the fact is that, if Edison wanted to do it now, he *would* have to have a licence. Today's genetic engineer is not free to choose his developmental path; those who designed the Abbeystead tunnel were not permitted to push a vent up into the cherished landscape that lay above it (nor did they have the institutional solidarity to refuse to build when this essential exercise of their engineering judgement was curtailed); the snail darter (that insignificant fish) trumped the Great American Dam; the Austrian electorate voted 51 per cent to 49 per cent against the commissioning of their vast (and vastly expensive) nuclear power station; the long-suffering residents of Ronan Point in the London Borough of Newham finally forced their local authority (and its expert advisors) to reverse its decisions (and the judgements of its advisors) and demolish these expensive but uninhabitable homes.

Those — the cranks, the romantics, the eco-freaks and the lumpen-proletariat — who, throughout the long years of the industrial age, could safely be ignored can no longer be ignored. That is the crucial, and perhaps none too palatable, message that the engineer now has to receive and respond to; that is the one great change that the transition to the Information Society (as it is sometimes called) has wrought in his environment; that is why he has to confront, and update, the internalised understanding that has for so long served him so well.

A slight shift of emphasis

The essential tension between the precise engineering science of the theorist and the much more judgemental engineering art of the hands-on practitioner has long been internalised by the engineering and design professions. In the particular risk field in which I happen to have some practical experience, this tension is nicely captured in the pithy comments of that great mountaineer, Don Whillans.

> There's no two ways about it: the Himalayas are very dangerous . . . People say its safe if you obey the rules, but they are wrong. If you have plenty of experience and understand the place then you have a chance of getting out OK but there are so many unexpected, unknown things that can happen. People call them freak conditions but they happen every few minutes. [Whillans 1984]

So here we have the risk assessor — going by the book and convinced that he can actually know the unknowable and anticipate the unanticipable — and the jackrabbit — cheerfully (but far from foolhardily) accepting the risks that the risk assessor has to insist are not there, and handling the unknowable and unanticipable through the exercise of the fine judgement that the continuous experience and understanding of those sorts of things develops.

But there is, of course, a third strategy. Unlike the risk assessor, a person may concede that freak conditions happen every few minutes and, unlike the jackrabbit, he may decide that those risks are simply not acceptable. This strategy (in mountaineering circles it is ascribed to the professional second — the person who holds the rope for the leader but never leads himself) is the one that the engineer used to be able to ignore. This is the risk-handling style that is not as well internalised as it will have to be. Where the engineer used to have to manage the tension between just two positions, he will now have to cope with a three-cornered tug-of-war. This is the shift of emphasis that he is going to have to make, and the sooner he makes it the less his loss of public esteem will be.

This third position is not easily reconciled with the other two. For all their differences, the risk assessor and the jackrabbit do both accept considerable risks; they differ only in their preferred methods of accepting and minimising them. But this third position declares unacceptable much that the other two find acceptable. As we move from the first two positions towards the third position, so benefits are dramatically transformed into penalties. Inevitably, if the engineer has to take account of this third position, he will have to concern himself more and more with the neglected question: what is a benefit? The debate used to be about the particular path along which a technology should be developed, now it is about whether or not we should have the technology itself.

Perhaps the most important consequence of this transition is that risk can no longer be separated out as a distinct thing to study and manage. In discussing risks we are now inevitably discussing the acceptability of the technologies that generate those risks. Safety engineering can no longer be detached from engineering proper. Culturally shaped attitudes to risk are nothing less than culturally shaped attitudes to technology. Risk studies are dead; long live the culture of technology!

The culture of technology

If engineering is indeed a mediating institution then we should expect these three positions not to be too clearly visible within it. Over the years many of the profession's crucial principles and approaches will have been unconsciously designed so as to straddle (and negotiate between) rather than to pinpoint (and give emphasis to) these divergent tendencies within the wider public. Even so, the kind of high-tech/white-heat-of-the-cutting-edge engineering that characterises much defence technology (so unlovingly described by Mary Kaldor in *The Baroque Arsenal*) can readily be contrasted with the cheap-and-cheerful style that often accompanies those smaller-scale enterprises that are more keenly exposed to the disciplines of the marketplace. And both can be contrasted with the kind of wilful minimalism that distinguishes those more recent arrivals on the engineering scene: the Appropriate Technologists.

In architecture (which is a design profession under whose skin I have, over the years, managed to creep) we have recently seen the 'grand design' dogmatism of the Modern Movement give way, on the one hand, to a cheap-and-cheerful style (of which Norman Foster is perhaps the ablest exponent) and, on the other hand, to the enormously laborious but scarcely visible modifications of the built environment carried out by Community Architects. Such persistent stylistic patterns, and their parallel shifts of emphasis, suggest that both engineering and architecture are, in fact, equipped with the internalised variety they will need if they are to cope with all the transitions that are currently upon them. If this is indeed the case then the social scientist would be ill-advised to try to teach these professional grandmothers how to suck eggs. Much better if he just helped them get on with it.*

Increasingly I find myself coming around to the view that if the social scientist is to communicate anything useful to the management of risk, and to the development of the technologies that underlie the risks that we

* But Charles Knevitt (1986) has suggested that the British architectural profession is about to fly apart under the strain. Though architecture is a 'broad church', he fears that it is not broad enough to take in the Community Architects (despite their royal patron, Prince Charles) and that they, like the civil engineers a century and a half ago, will have to set up on their own. The sort of self-awareness that the anthropologist can bring to the design professions would, I believe, alert the Royal Institute of British Architects to this sort of risk and encourage them to broaden their church enough to avert the breakaway.

manage, then he will have to learn to speak the strange languages of the tribes that actually do the dirty work: the engineers, the architects, the civil servants, the lawyers and so on. Since all (I suspect) are skilled in institutionalised mediation, little purpose will be served by the social scientist's re-invention of that particular wheel. Rather, he should concern himself with the care and repair of the wheels that already exist. This is an urgent and far from trivial task. On the one hand, it requires engineers to regain their humane and civilising breadth. On the other hand, it requires social scientists to concede that much hard-won wisdom is centrally encoded in long-established ways of doing things that are not readily accessible to them. The longer each delays the less chance there is of them both succeeding.

References

Chilver, Henry, 1975, 'Wider Implications of Catastrophe Theory', *Nature*, **254**.

Clark, William, C., 1980, 'Witches, Floods and Wonder Drugs', in Schwing, R. and Albers, W., (eds), *Societal Risk Assessment*, Plenum, New York.

Critchley, Octavius H., 1978, 'Aspects of the Historical Philosophical and Mathematical Background to the Management of Nuclear Power Plants in the United Kingdom', *Radiation Protection in Nuclear Power Plants and the Fuel Cycle*, The British Nuclear Energy Society, London, 11–18.

Grauer, M., Thompson, M. and Wierzbicki, A. (eds), 1975, *Plural Rationality and Interactive Decision Processes*, Lecture notes in Economics and Mathematical Systems, No 248, Springer, Berlin.

Hapgood, Fred, 1979, 'Risk-benefit Analysis: Putting a Price on Life', *The Atlantic*.

Keynes, J.M., 1973, *Treatise on Probability*, Macmillan, London.

Knevitt, Charles, 1986, 'The Architect as Anti-hero', *The Times*, 7 February.

Knight, F.H., 1921, *Risk, Uncertainty and Profit*, republished 1965, Harper, New York.

Muir Wood, A.M., 1980, letter to *The Times*, 7 February.

Rasmussen, N.C. *et al*, 1975, *Reactor Safety Study, WASH 1400*, US Nuclear Regulatory Commission (NUREG-75/014). Washington DC.

Rothschild, Lord, 1978, 'Risk', *The Listener*, London, November.

Silcock, Bryan, 1979, 'Risky Business', *The Sunday Times*, London, 30 December.

Slovic, P., Fischhoff, B. and Lichtenstein, S., 1979, 'Which Risks are Acceptable?' *Environment*, **21** (4).

Thompson, Hunter S., 1980, *The Great Shark Hunt*, Picador/Pan, London.

Thompson, Michael, 1980, 'The Aesthetics of Risk: Culture or Context?' in Schwing, R. and Albers, W. (eds), *Societal Risk Assessment*, Plenum, New York.

Thompson, Michael, (forthcoming) 'Socially Viable Ideas of Nature', in Baark, E. and Svedin, V. (eds), *Nature, Culture, Technology: Towards a New Conceptual Framework*, Macmillan, London.

Whillans, Don, 1984, 'As reported, on the occasion of Whillans's fiftieth birthday, by Ronald Faux in an article, 'The Only Way is Up', *The Times*.

Index